Springer Theses

Recognizing Outstanding Ph.D. Research

Aims and Scope

The series "Springer Theses" brings together a selection of the very best Ph.D. theses from around the world and across the physical sciences. Nominated and endorsed by two recognized specialists, each published volume has been selected for its scientific excellence and the high impact of its contents for the pertinent field of research. For greater accessibility to non-specialists, the published versions include an extended introduction, as well as a foreword by the student's supervisor explaining the special relevance of the work for the field. As a whole, the series will provide a valuable resource both for newcomers to the research fields described, and for other scientists seeking detailed background information on special questions. Finally, it provides an accredited documentation of the valuable contributions made by today's younger generation of scientists.

Theses may be nominated for publication in this series by heads of department at internationally leading universities or institutes and should fulfill all of the following criteria

- They must be written in good English.
- The topic should fall within the confines of Chemistry, Physics, Earth Sciences, Engineering and related interdisciplinary fields such as Materials, Nanoscience, Chemical Engineering, Complex Systems and Biophysics.
- The work reported in the thesis must represent a significant scientific advance.
- If the thesis includes previously published material, permission to reproduce this must be gained from the respective copyright holder (a maximum 30% of the thesis should be a verbatim reproduction from the author's previous publications).
- They must have been examined and passed during the 12 months prior to nomination.
- Each thesis should include a foreword by the supervisor outlining the significance of its content.
- The theses should have a clearly defined structure including an introduction accessible to new PhD students and scientists not expert in the relevant field.

Indexed by zbMATH.

Toni Mäkelä

Towards Global Interpretation of LHC Data

SM and EFT Couplings from Jet and Top-Quark Measurements at CMS

Doctoral Thesis accepted by
The University of Hamburg, Hamburg, Germany

Springer

Author
Dr. Toni Mäkelä
BP2 Theoretical Physics Division
National Centre for Nuclear Research
Warsaw, Poland

Supervisor
Prof. Dr. Katerina Lipka
DESY
Hamburg, Germany

University of Wuppertal
Wuppertal, Germany

ISSN 2190-5053 ISSN 2190-5061 (electronic)
Springer Theses
ISBN 978-3-031-29778-6 ISBN 978-3-031-29779-3 (eBook)
https://doi.org/10.1007/978-3-031-29779-3

This Springer imprint is published by the registered company Springer Nature Switzerland AG
The registered company address is: Gewerbestrasse 11, 6330 Cham, Switzerland

Thus ought man to be an impersonation of the divine process of nature, and to show forth the union of the infinite with the finite.

—James Clerk Maxwell

Supervisor's Foreword

The elementary structure and interactions of matter are described by the standard model of particle physics, a quantum field theory based on gauge symmetries and operating with masses of particles and the interaction couplings as fundamental parameters. Similar to the periodic system in chemistry, the standard model takes the central role in the description of the subatomic world. In spite of the great success of this model to describe the observed phenomena, some aspects of its structure and relations of its parameters hint at the presence of new physics—yet unknown fundamental interactions, setting up at very large energy scales. For example, the standard model sorts the fundamental building blocks of matter, the quarks, into three families. However, the stable nuclear matter—everything we can see and touch—is built from the members of the first family only. This fact is not explained yet. One hypothesis suggests that the known quarks could have a structure themselves and be subject to a yet unknown interaction setting in at the very high energy scale, inaccessible by direct measurements in existing experiments. To date, the most promising indirect searches for such new physics are performed by studying the quark interactions at the Large Hadron Collider, LHC, at CERN. There, the strongly interacting particles called hadrons, in particular protons, are brought to collisions at very high energies.

The main challenge of studying quark interactions in hadron collisions is the fact that quarks are confined inside hadrons by the strong force and cannot be observed as free particles. Once kicked out of a proton, quarks form collimated sprays of hadrons, called jets. By measuring the energy and the direction of a jet, one can conclude on the properties of the original quark. Each deviation of the experimentally measured jet rates from the standard model predictions could be a signal of new physics. The new physics contributions are usually described by the effective field theory.

The observed jets may emerge from a new interaction, as well as from the strong interaction of the standard model. Therefore, the interpretation of the measurements relies to large extent on the precision of the standard model parameters and of the distributions of quarks and gluons in the proton. This problem persisted since the first searches for new physics using jets in proton-antiproton collisions, performed in the 1990s at the Tevatron in Fermilab, demanding a novel approach to extract the couplings of the new physics, unbiased from the strong interaction of the standard model. This long-standing issue is resolved in the thesis by Toni Mäkelä, who has developed a new analysis method for the simultaneous extraction of the standard model parameters and the couplings of a possible new interaction. In this thesis, the standard model prediction of jet rates in proton-proton collisions at the LHC is extended to account for new physics contributions from so-called quark contact interactions, provided by effective field theory. The resulting prediction is confronted with the most recent measurement of inclusive jet and top quark-antiquark pair production at the LHC at the center of mass energy of 13 TeV, collected by the CMS experiment. Together with the CMS measurements, the data of the electron-proton collisions at HERA accelerator at DESY are used, which serve as a basis for any determination of the quark and gluon distributions in the proton.

In the thesis, for the first time, the data of HERA and LHC are used to simultaneously constrain the minimal energy at which the novel interaction can emerge, together with the quark and gluon distributions in the proton, and the values of the top quark mass and strong coupling constant. As a result, the most stringent constraints on the couplings of the new interaction are imposed, and the best precision is achieved for the values of the top quark pole mass and of the strong coupling constant, to date. The latter is determined with an accuracy of about 1.5%, which is currently the most precise single measurement at a hadron collider.

Hamburg, Germany
December 2022

Prof. Dr. Katerina Lipka

Abstract

This thesis presents the first global interpretation of the measurements of double-differential cross sections of inclusive jet and top quark-antiquark pair ($\mathrm{t\bar{t}}$) production in proton-proton collisions at the Large Hadron Collider (LHC) at the center of mass energy of $\sqrt{s} = 13$ TeV, together with inclusive measurements in Deep Inelastic electron-proton Scattering (DIS) at HERA. The LHC data have been recorded by the Compact Muon Solenoid (CMS) experiment in 2016 and correspond to the integrated luminosity of 33.5 fb^{-1} for inclusive jet measurement or 35.9 fb^{-1} for the $\mathrm{t\bar{t}}$ measurement.

This novel analysis is performed at next-to-leading order (NLO) of perturbative quantum chromodynamics (QCD), where the inclusive neutral and charged current DIS measurements at HERA, the CMS jet measurements and the normalized triple differential cross sections of $\mathrm{t\bar{t}}$ production are used together to constrain the parameters of QCD and impose constraints on new physics. In particular, the Standard Model Lagrangian is extended to incorporate effective contributions from quark contact interactions (CI). In this analysis, the constraints on the parton distributions, the strong coupling constant $\alpha_S\,(m_Z)$ and the top quark pole mass m_t^{pole} are obtained, and simultaneously, limits on the scale Λ of the CI are imposed. This results in $\alpha_S(m_Z) = 0.1188 \pm 0.0031$, $m_t^{\mathrm{pole}} = 170.4 \pm 0.7$ GeV and the 95% confidence level limits of $\Lambda > 24$ TeV for purely left-handed, $\Lambda > 32$ TeV for vector-like and $\Lambda > 31$ TeV for axial vector-like exchanges. For the first time, the LHC data are used for extracting the Wilson coefficient of the CI, unbiased from the assumptions on the Standard Model parameters.

The jet and top quark production at the LHC are further investigated individually, to determine the QCD parameters at the highest precision. A Standard Model analysis is performed at next-to-next-to-leading order (NNLO) of perturbative QCD using the CMS jet cross section measurements together with the HERA DIS data. The parton distribution functions in the proton are extracted simultaneously with the strong coupling constant $\alpha_S\,(m_Z) = 0.1166 \pm 0.0017$. This is the most precise single measurement of the strong coupling constant at the LHC, to date.

Further, the CMS measurements of the $t\bar{t}$ cross sections are used to extract the top quark mass in the MSR renormalisation scheme. In this scheme, the mass dependence in the difference between the pole and modified minimal subtraction ($\overline{\text{MS}}$) masses is replaced by a scale parameter R. The first study of the behaviour of the R scale is presented. The extracted top quark MSR mass obtained using CMS measurements results in $m_t^{\text{MSRn}}(R = 80\ \text{GeV}) = 169.3^{+0.6}_{-0.7}$ GeV, which is up to now the most precise value of the top quark MSRn mass extracted from experimental data.

The presented investigations of the fundamental couplings of the Standard Model, and of new physics, are at the edge of precision in experimental physics and high energy phenomenology, paving the road towards a global interpretation of LHC data.

Publications Related to this Thesis

- T. Mäkelä, A. Hoang, K. Lipka, S.O. Moch. "Investigation of the scale dependence in the MSR and $\overline{\text{MS}}$ top quark mass schemes for the $t\bar{t}$ invariant mass differential cross section using LHC data". January 2023. arXiv: 2301.03546 [hep-ph]
- CMS Collaboration. "Measurement and QCD analysis of double-differential inclusive jet cross sections in pp collisions at $\sqrt{s}$ 13 TeV", *JHEP* 02 (2022), p. 142. http://dx.doi.org/10.1007/JHEP02(2022)142
- T. Mäkelä, A. Hoang, K. Lipka, S.O. Moch. "Single-differential top quark pair production cross sections with running mass schemes at NLO", arXiv: 2211.13607 [hep-ph]
- H. Abdolmaleki et al. "xFitter: An Open Source QCD Analysis Framework. A resource and reference document for the Snowmass study". arXiv: 2206.12465 [hep-ph]
- D. d'Enterria et al. "The strong coupling constant: State of the art and the decade ahead". arXiv: 2203.08271 [hep-ph]
- T. Mäkelä [for the CMS Collaboration]. "Precision QCD measurements from CMS". arXiv: 2206.11624 [hep-ph]

Acknowledgements

Knowing now in hindsight the effort that a Ph.D. project requires, I first and foremost wish to express my gratitude to my supervisor Prof. Dr. Katerina Lipka, and to Prof. Dr. Sven-Olaf Moch, for the work we carried out together, and for everything I learnt about high energy physics as well as research and life in academia. PD. Dr. Alexander Glazov deserves special thanks for sharing his priceless insight on using and extending the xFITTER framework. I also thank Prof. Dr. Dieter Horns for chairing the defense, and Prof. Dr. Gregor Kasieczka for being a part of the committee.

All this wouldn't have been possible without the helpful discussions I had with many colleagues. In particular, I wish to thank Dr. Patrick Connor, Dr. Matteo Defranchis, Dr. Engin Eren, Dr. Jun Gao, Prof. Dr. André Hoang, Dr. Svenja Pflitsch, Prof. Dr. Harrison Prosper, PD. Dr. Klaus Rabbertz and Dr. Oleksandr Zenaiev.

My fellow group members, office inhabitants, other students, etc. at DESY and Universität Hamburg deserve to be thanked for the good times in- and outside of the campus. Of those not already mentioned, thanks to Ali, Ana, Bakar, Beatriz, Evan, Fang-Ying, Federico, Josry, Qun, Sam, Sebastian, Simone, Valentina and Vitaly, and apologies to anyone I might have forgotten. However, since a large part of my time as a Ph.D. student coincided with the COVID-19 pandemic, I'd like to thank Antti, Caro, Frauke, Tony, Ville and the progressive minds at Retrogramofoni + Ilkka and The Boys for keeping in touch virtually. I remain especially indebted to Ted for being there with me in Hamburg through it all. 梁先輩の有意義な教訓を忘れません。友達になってくれてありがとう!

I wish to thank my family for always being there for me, regardless of the spatial distance. Kiitokset tuestanne ja ymmärryksestänne sekä siitä, että piditte mielessäni mikä elämässä on tärkeää.

Contents

Chapter 1
Introduction

The current understanding of the dynamics between elementary particles is encoded into a quantum field theory (QFT) known as the Standard Model (SM) of particle physics. So far, the SM has described the observations in particle physics experiments with great success. However, it is not anticipated to remain the final theory of particle physics due to its well-known shortcomings. While the SM predicts the relations of its parameters, which include the couplings of the interactions and the masses of elementary particles, their values need to be determined experimentally. Furthermore, it is unexplained why the SM operates with 3 fermion families, although only one is needed for building stable matter, or why there are large differences in the masses of the particles within a family. It is also possible that the electroweak vacuum is not stable in the SM, but resolving whether or not this is the case requires increasing precision in the measurements of the top quark mass m_{t} and the value of the strong coupling constant α_S. Both α_S and m_{t} are measured with extreme precision in the context of this thesis, along with investigating the phenomenological aspects of their determination—individually, using cutting-edge phenomenology for jet and $\mathrm{t\bar{t}}$ production, but also together, exploring the sensitivity of both processes to the gluon distribution and α_S in proton-proton collisions.

There is an abundance of theories beyond the Standard Model (BSM), but so far none have been confirmed experimentally. The extensive searches carried out at collider experiments have lead to increasing exclusion limits for the scale of new physics, particularly in proton-proton collisions at the Large Hadron Collider (LHC) at CERN. With the most recently accomplished Run 2 at a center-of mass-energy of 13 TeV, LHC is the facility at the very frontier of collision energy.

The aim of a *direct search* for new physics is the observation of a resonance peak in the spectrum of an observable, such as a cross section. However, there are hints that a direct observation of new physics, a yet unknown very massive elementary particle or a fundamentally new interaction, is out of reach even at the LHC. Moreover, the SM assumes elementary particles to be pointlike, and experimental limits of $\mathcal{O}(10^{-17})$ cm have been set for the quark radii. If the radii are fundamentally finite, resulting

T. Mäkelä, *Towards Global Interpretation of LHC Data*, Springer Theses,
https://doi.org/10.1007/978-3-031-29779-3_1

for instance from more elementary particles constituting the quarks as composite particles, it would appear as a deviation from the SM cross section spectra [1, 2]. If the BSM effects are not expected to be seen as resonance peaks, but as a continuously increasing deviation from the SM prediction as a function of energy, the search is referred to as *indirect*. A historical example of a successful indirect search is provided by the observation of the continuous energy spectrum in beta decay, which led to the formulation of the electroweak theory.

The contributions of new physics are mostly approximated by the Standard Model effective field theory (SMEFT), which extends the SM by operators that describe new interactions between SM particles. The cause of the interaction, e.g. the exchange of new particles or substructure of SM particles, is integrated out, and predictions at energies below the BSM scale are obtained without the need to resolve the exact mechanism. However, achieving sensitivity to minute differences in the spectra of physical observables requires both the measurements and the predictions to be extremely precise.

In the SM, the dominant processes in the proton-proton collisions at the LHC are described by quantum chromodynamics (QCD), and the most fundamental process for studying QCD is the production of *jets*: since quarks or gluons (together called partons) cannot be observed as free particles due to the confinement in QCD, the partons kicked out of a proton in a pp collision form collimated sprays of color-neutral hadrons, so-called jets, which preserve the momentum and direction of the original parton. Jet production thus sheds light on the parton distribution functions (PDFs) in the proton, is directly sensitive to the value of the strong coupling constant α_S, and at the same time probes the scale of new physics at high transverse momenta. In particular, the PDFs and the value of α_S play a central role in the interpretation of the LHC measurements.

The PDFs are functions of the energy scale at which the proton structure is probed and of the partonic fraction x of the proton momentum. While the energy scale dependence of the PDFs is known in perturbative QCD, the x-dependence cannot yet be calculated. Therefore, the PDFs need to be extracted using experimental data. The PDF determination assumes the validity of the SM also at high transverse momenta where new physics can emerge, possibly absorbing the new physics inside the proton structure and therefore causing a bias in the interpretation of the measurements in an indirect search. This problem is assessed in this thesis by following an unbiased analysis strategy in a search for 4-quark contact interactions (CI).

The main focus of the thesis is the comprehensive QCD analysis of the double-differential inclusive jet data measured by the CMS Collaboration in pp collisions at $\sqrt{s} = 13$ TeV, corresponding to an integrated luminosity of 33.5 fb^{-1}. The jets are reconstructed using the anti-k_T algorithm with the distance parameter 0.7. In an analysis at next-to-next-to-leading order (NNLO) in SM QCD, these data are used together with the HERA inclusive deep inelastic scattering (DIS) cross sections [3], resulting in a simultaneous extraction of the PDFs and $\alpha_S(m_Z)$. The obtained NNLO value of $\alpha_S(m_Z)$ is the most precise value extracted in a single experiment at the LHC, to date. In an analysis performed at next-to-leading order (NLO), also the CMS triple-differential top quark-antiquark production cross section measurements at $\sqrt{s} = 13$ TeV are used, giving additional sensitivity to the top quark mass. The NLO analysis is performed using SM or, alternatively, SMEFT predictions for the jet cross section, leading to a first-ever simultaneous extraction of the PDFs, $\alpha_S(m_Z)$, the top quark mass and the Wilson coefficient of 4-quark CI using LHC data.

An important topic of the phenomenology of top quark physics investigated in the thesis is the renormalization of the top quark mass. Vacuum polarization requires the renormalization of all charges and couplings in a theory, and therefore also the masses of quarks become dependent on the choice of renormalization scheme. Several renormalization schemes and in turn definitions for a top quark mass have been suggested, and have different advantages and disadvantages with respect to the interpretation of experimental measurements in terms of the top quark mass. For example, the difference between the values of the top quark mass in the widely used pole and modified minimal subtraction ($\overline{\mathrm{MS}}$) mass $\overline{m}_t(\overline{m}_t)$ can be of the order of several GeV [4].

The pole mass scheme is the perturbative QCD equivalent of the on-shell mass of a free particle. It redefines the quark mass parameter order-by-order in perturbation theory, making it correspond to the pole of the quark propagator. However, the quark pole masses are affected fundamentally by an infrared sensitivity known as the renormalon ambiguity, which is of the order of the QCD scale Λ_{QCD} [5–7]. The $\overline{\mathrm{MS}}$ mass is an example of a *short-distance* mass not suffering from such ambiguities, and is a convenient choice for loop calculations of mass-dependent inclusive physical observables dominated by short-distance effects [8]. However, such masses depend on the *renormalization scale* at which the infinities are subtracted. This scale dependence is known as *running*, and is described by a renormalization group equation (RGE).

A way to bridge the pole and $\overline{\mathrm{MS}}$ masses is offered by the MSR scheme [9]. The MSR mass is matched to the $\overline{\mathrm{MS}}$ mass $\overline{m}_t$ at a mass renormalization scale $R = \overline{m}_t(\overline{m}_t)$, and it approaches that of the pole mass scheme as $R \to 0$. This thesis discusses the implementation of the MSR scheme into the inclusive top quark-antiquark pair production cross section calculator HATHOR [10] and the single-differential $\mathrm{t\bar{t}}$ production cross section computation in the MCFM Monte Carlo program version 6.8 [11–13]. These allow the first examination of the behavior of the scale R and the properties of the total and differential $\mathrm{t\bar{t}}$ cross sections using the MSR scheme. In this thesis, the top quark MSR mass is determined from the $\mathrm{t\bar{t}}$ pair production cross section measured as a function of the top quark-antiquark pair invariant mass $m_{\mathrm{t\bar{t}}}$.

The structure of the thesis is as follows. A theoretical overview of the SM of particle physics, renormalization and proton structure is given in Chap. 2, focusing on the role of jets and the top quark in the SM. The approach of effective theories is discussed in Chap. 3. The LHC and the CMS Experiment are sketched in Chap. 4. Chapter 5 contains the discussion of the implementation of the MSR renormalization scheme to the HATHOR and MCFM programs, and the studies of the mass renormalization scale, independent of the QCD factorization and renormalization scales, as well as the extraction of the top quark MSR mass from CMS data. Chapter 6 details the XFITTER open QCD analysis framework, which is used for the QCD analysis of the CMS 13 TeV inclusive jet and $t\bar{t}$ production cross section data presented in Chap. 7. The thesis is summarized and concluded in Chap. 8.

The *Einstein summation* convention is adopted throughout the thesis whenever indices are repeated up and down, so that $v_\mu v^\mu \equiv (v_0)^2 - \sum_{i=1}^{3}(v_i)^2$ for a 4-vector v, and $u_i u^i \equiv \sum_{i=1}(u_i)^2$ when i is not a spacetime index. *Natural units* are employed, i.e. the speed of light and Planck's reduced constant $\hbar$ are set to 1. Thus the units of mass and energy agree, and the basic unit is the electronvolt (eV). Further quantities' units can be obtained as powers of the unit of energy, and the power is referred to as the quantity's *mass dimension*. Greek letters are used for spacetime indices, whereas other indices are indicated by Roman letters. The Lagrangian density $\mathscr{L}$ is referred to as the Lagrangian.

References

1. Kopp G, Schaile D, Spira M, Zerwas PM (1995) Bounds on radii and magnetic dipole moments of quarks and leptons from LEP. SLC HERA Z Phys C 65:545–550
2. H1 and ZEUS Collaborations (2015) Combination of measurements of inclusive deep inelastic e±p scattering cross sections and QCD analysis of HERA data. Eur Phys J C 75:580
3. Abramowicz H et al (2016) Limits on the effective quark radius from inclusive ep scattering at HERA. Phys Lett B 757:468–472
4. Langenfeld U, Moch S, Uwer P (2009) Measuring the running top-quark mass. Phys Rev D 80:054009
5. Bigi IIY, Shifman MA, Uraltsev NG, Vainshtein AI (1994) The Pole mass of the heavy quark: perturbation theory and beyond. Phys Rev D 50:2234–2246
6. Beneke M, Braun VM (1994) Heavy quark effective theory beyond perturbation theory: Renormalons, the pole mass and the residual mass term. Nucl Phys B 426:301–343
7. Smith MC, Willenbrock SS (1997) Top quark pole mass. Phys Rev Lett 79:3825–3828
8. Chetyrkin KG, Steinhauser M (1999) Short distance mass of a heavy quark at order α_s^3. Phys Rev Lett 83:4001–4004
9. Hoang AH, Stewart IW (2008) Top mass measurements from jets and the Tevatron top-quark mass. Nucl Phys B Proc Suppl 185:220–226 (Bellettini G, Chiarelli G, Tenchini R (eds))
10. Aliev M et al (2011) HATHOR: HAdronic Top and Heavy quarks crOss section calculatoR. Comput Phys Commun 182:1034–1046
11. Campbell JM, Ellis RK, Giele WT (2015) A multi-threaded version of MCFM. Eur Phys J C 75:246
12. Campbell JM, Ellis RK (1999) An update on vector boson pair production at hadron colliders. Phys Rev D 60:113006
13. Campbell JM, Ellis RK, Williams C (2011) Vector boson pair production at the LHC. JHEP 07:018

Chapter 2
The Role of Jets and the Top Quark in the Standard Model

This chapter presents an overview of the Standard Model in Sect. 2.1. The issue of the stability of the electroweak vacuum is presented in Sect. 2.2, before turning to more details on quantum choromodynamics in Sect. 2.3. A general discussion of the role of renormalization in quantum field theory is given in Sect. 2.4. The application of renormalization group equations in Standard Model physics is discussed using the example of the running of the strong coupling constant in Sect. 2.5, and the definition and renormalization of quark masses with an emphasis on the top quark in Sect. 2.6. Phenomenological aspects of jet and top quark production are covered in Sect. 2.7, and the structure of the proton in Sect. 2.8.

2.1 Overview of the Standard Model

The Standard Model (SM) provides a quantum field theory (QFT) explanation for electromagnetism, as well as the strong and weak forces, via a combination of quantum electrodynamics (QED), the electroweak (EW) theory and quantum chromodynamics (QCD). As a Lorentz invariant theory, the SM assumes CPT invariance: the theory remains unchanged under a simultaneous reversal of charges (C), parity (P) and time (T). However, no individual one or a combination of two of them are required symmetries in the SM.

Another central feature of the SM is *gauge invariance* under the symmetry group $SU(3) \times SU(2) \times U(1)$. Matter particles arise from excitations of fermionic fields ψ, which change under a gauge transformation as

$$\psi \to \exp\left(it^n \theta_n\right) \psi, \tag{2.1}$$

T. Mäkelä, *Towards Global Interpretation of LHC Data*, Springer Theses,
https://doi.org/10.1007/978-3-031-29779-3_2

where t^n are the N generators of $SU(N)$. If the parameters θ_n are independent of the space-time coordinates, the transformation is *global*, otherwise the transformation is *local*. The gauge is not a physical quantity, but must be fixed in the process of computing theoretical predictions for observables. The motivation for requiring gauge invariance is that an unphysical choice should not affect physical quantities, and failure to preserve it can lead to non-unitary evolution operators. The controlled way of breaking gauge invariance by the introduction of covariant gauges is of key importance in proofs of renormalizability [1].

The kinetic terms of Lagrangians involve derivatives of the fields, but these are not invariant under local gauge transformations. This is cured by a replacement of the form

$$\mathbb{1}\partial_\mu \rightarrow \mathbb{1}\partial_\mu - igt_n A^n_\mu, \tag{2.2}$$

introducing N *gauge boson fields* with coupling strengths g and components A^n_μ, one for each generator t_n. The gauge fields then transform as [1]

$$A^n_\mu \rightarrow A^n_\mu + \frac{1}{g}\partial_\mu \theta^n - f^n_{\ ab}\theta^a A^b_\mu, \tag{2.3}$$

where $f^n_{\ ab}$ are the structure constants of $SU(N)$. The eight generators of $SU(3)$ then introduce the eight gluon fields G^a_μ, $SU(2)$ generates the three fields W^i_μ, and $U(1)$ induces the field B_μ. The mixing of the fields W^i_μ and B_μ in the electroweak spontaneous symmetry breaking (EWSSB) gives rise to the 3 massive electroweak bosons W^+, W^-, Z and a massless photon γ. The *gauge covariant derivatives* needed for the SM Lagrangian read

$$\begin{cases} D_\mu &= \partial_\mu - \frac{i}{2} g_1 Y B_\mu - \frac{i}{2} g_2 \sigma_i W^i_\mu \\ D'_\mu &= D_\mu - \frac{i}{2} g_3 \lambda_a G^a_\mu \end{cases}, \tag{2.4}$$

where Y is the hypercharge operator, σ_i are the Pauli matrices and λ_i the Gell-Mann matrices. The $SU(3)$ coupling is usually rewritten in terms of the *coupling constant* $\alpha_S \equiv g_3^2/(4\pi)$, whereas the couplings g_1 and g_2 can be expressed in terms of other SM parameters.

The SM Lagrangian is given by [2]

$$\begin{aligned} \mathscr{L}_{\mathrm{SM}} = -\frac{1}{4}\Big(& \underbrace{F_{\mu\nu}F^{\mu\nu}}_{\substack{U(1)\text{ gauge}\\ \text{sector}}} + \underbrace{W^i_{\mu\nu}W^{\mu\nu}_i}_{\substack{SU(2)\text{ gauge}\\ \text{sector}}} + \underbrace{G^a_{\mu\nu}G^{\mu\nu}_a}_{\substack{SU(3)\text{ gauge}\\ \text{sector}}}\Big) \\ & \underbrace{+ i\overline{L}^i_L \gamma^\mu D_\mu L_{Li} + i\overline{\ell}^i_R \gamma^\mu D_\mu \ell_{Ri}}_{\text{Lepton kinetic sector}} \\ & \underbrace{+ i\overline{Q}^i_L \gamma^\mu D'_\mu Q_{Li} + i\overline{u}^i_R \gamma^\mu D'_\mu u_{Ri} + i\overline{d}^i_R \gamma^\mu D'_\mu d_{Ri}}_{\text{Quark kinetic sector}} \end{aligned}$$

$$- \underbrace{\left(\overline{Q^i_L} Y^{(u)}_{ij} H u^j_R + \overline{Q^i_L} Y^{(d)}_{ij} H d^j_R + \overline{L^i_L} Y^{(l)}_{ij} H \ell^j_R + \text{h.c.} \right)}_{\text{Yukawa sector}}$$
$$+ \underbrace{(D^\mu H)^\dagger (D_\mu H)}_{\text{Higgs kinetic sector}} - \underbrace{\left(\mu^2 H^\dagger H + \lambda (H^\dagger H)^2 \right)}_{\text{Higgs potential } V(H)}, \tag{2.5}$$

where h.c. stands for the Hermitian conjugate, γ^μ are the Dirac matrices, and the subscripts L, R denote left- and right handed fields, respectively. H is the Higgs doublet, and Q^i_L (L^i_L) are the left-handed quark (lepton) doublets, while u^i_R (d^i_R) stand for the right-handed up (down) type quarks, and ℓ^i_L is for the left-handed leptons. The *field strength tensors* for the Abelian $U(1)$ and the non-Abelian $SU(2)$, $SU(3)$ read

$$U(1): \; F_{\mu\nu} = \partial_\mu B_\nu - \partial_\nu B_\mu \tag{2.6}$$
$$SU(2): \; W^i_{\mu\nu} = \partial_\mu W^i_\nu - \partial_\nu W^i_\mu + g_2 \epsilon^i{}_{jk} W^j_\mu W^k_\nu, \tag{2.7}$$
$$SU(3): \; G^a_{\mu\nu} = \partial_\mu G^a_\nu - \partial_\nu G^a_\mu + g_3 f^a{}_{bc} G^b_\mu G^c_\nu, \tag{2.8}$$

where $\epsilon^i{}_{jk}$ and $f^a{}_{bc}$ are the structure constants of $SU(2)$ and $SU(3)$, respectively. The trace of the product of a field strength tensor with itself is gauge invariant, and serves to introduce the kinetic terms for all gauge bosons. Further, it gives the trilinear and quartic interaction vertices for the weak bosons and gluons, but no self-interaction vertices for the photons.

In principle, the theory also allows for the gauge sector like term

$$\theta \frac{g^2}{64\pi^2} G^a_{\mu\nu} \epsilon_{\mu\nu\rho\sigma} G^{\rho\sigma}_a, \tag{2.9}$$

where $\epsilon_{\mu\nu\rho\sigma}$ is the total antisymmetric tensor. The term does not necessarily vanish for $SU(3)$, and introducing the θ term would violate CP symmetry [3], which has never been observed in QCD. Consequently, θ is estimated to be close to zero [4]. However, the SM has a priori no requirement for θ to be small, and this is referred to as the *strong CP problem*. A possible explanation has been offered by the Peccei-Quinn theory [5, 6], in which the CP violation is suppressed by a new pseudoscalar particle known as the *axion* [7, 8]. However, heavy axions solving the CP problem have been ruled out by experiments such as Ref. [9], and further models with decay constant values yet to be excluded have been developed [10–13].

The quark and lepton kinetic sectors of Eq. (2.5) contain the kinetic terms for the fermions in the form of the Dirac Lagrangian, and the covariant derivative introduces their couplings to the gauge bosons. The fermionic fields of the SM are divided, depending on which bosons they are coupled to, into the leptons

$$L_L = \left(\begin{bmatrix} \nu_{eL} \\ e_L \end{bmatrix}, \begin{bmatrix} \nu_{\mu L} \\ \mu_L \end{bmatrix}, \begin{bmatrix} \nu_{\tau L} \\ \tau_L \end{bmatrix} \right), \; \ell_R = (e_R, \mu_R, \tau_R) \tag{2.10}$$

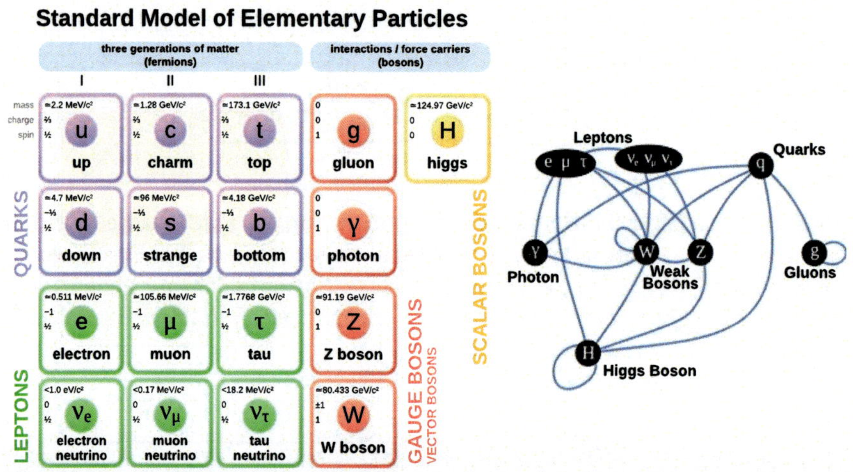

Fig. 2.1 *Left*: the elementary particle content of the SM. Image taken from Ref. [14]. *Right*: a schematic of the couplings between SM particles. Image taken from Ref. [15]

and the quarks

$$Q_L = \left(\begin{bmatrix}\mathrm{u}_L\\\mathrm{d}_L\end{bmatrix}, \begin{bmatrix}\mathrm{c}_L\\\mathrm{s}_L\end{bmatrix}, \begin{bmatrix}\mathrm{t}_L\\\mathrm{b}_L\end{bmatrix}\right), \; u_R = (\mathrm{u}_R, \mathrm{c}_R, \mathrm{t}_R), \; d_R = (\mathrm{d}_R, \mathrm{s}_R, \mathrm{b}_R). \tag{2.11}$$

The couplings are summarized together with the particle content of the SM in Fig. 2.1.

In the form of Eq. (2.5), the SM Lagrangian contains no explicit mass terms, since they are prohibited by gauge invariance [1]. The masses are only introduced via the Brout-Englert-Higgs mechanism [16, 17], in which the Higgs potential takes the form illustrated in Fig. 2.2 when $\mu^2 < 0$ and $\lambda > 0$. The configuration where both components of the Higgs doublet H are at zero is an unstable extremum, where only the gauge and kinetic sectors have non-vanishing contributions to $\mathscr{L}_{\mathrm{SM}}$. In the EWSSB, the Higgs field collapses into a state of lower potential, acquiring a vacuum expectation value $v = |\mu|\lambda\sqrt{2}$ so that

$$H = \begin{bmatrix}H_1\\H_2\end{bmatrix} \rightarrow \frac{1}{\sqrt{2}}\begin{bmatrix}0\\v+h\end{bmatrix}, \tag{2.12}$$

where the first component of H can always be fixed to zero; changes of the Higgs field in this direction correspond to unphysical *Goldstone bosons*. In contrast, the excitations h about the new vacuum value manifest as a physical scalar, the Higgs boson, which was observed at the LHC by the ATLAS [18] and CMS Collaborations [19].

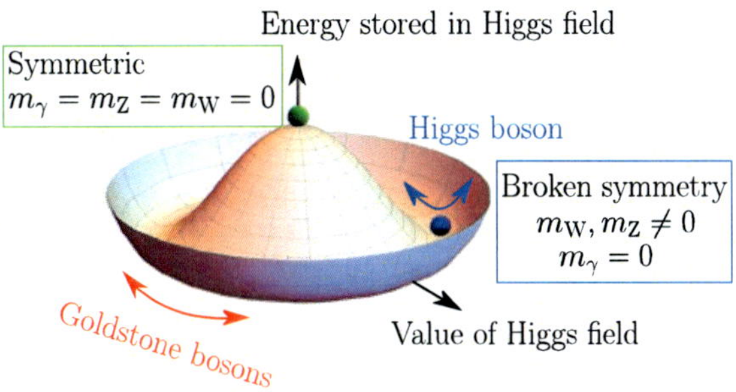

Fig. 2.2 The Higgs mechanism. Image after Ref. [20]

As the EW symmetry breaks, the Higgs kinetic term in Eq. (2.5) introduces the mass terms and couplings to h for the weak bosons via the non-zero vacuum expectation value, while the fermion mass terms and Higgs couplings arise from the Yukawa sector. However, the Yukawa matrices Y_{ij} are general complex matrices and not necessarily Hermitian. Obtaining real diagonal mass matrices requires a change from the original *flavor basis* to *mass eigenstates*. This process also affects the quark kinetic terms, leading to flavor mixing in the interactions between quarks and $W^{\pm}$ bosons. The mixing is described by the Cabibbo-Kobayashi-Maskawa (CKM) matrix [1]

$$V_{\text{CKM}} = \begin{bmatrix} V_{\text{ud}} & V_{\text{us}} & V_{\text{ub}} \\ V_{\text{cd}} & V_{\text{cs}} & V_{\text{cb}} \\ V_{\text{td}} & V_{\text{ts}} & V_{\text{tb}} \end{bmatrix}. \tag{2.13}$$

As a unitary complex matrix, V has nine degrees of freedom: three angles and six phases. However, five of the phases can be set to zero by suitable transformations of the quark operators [1]. Hence, the CKM matrix is presented in terms of the sines (s_{ij}) and cosines (c_{ij}) of three angles θ_{12}, θ_{13}, θ_{23}, and one phase δ as the product [4]

$$V_{\text{CKM}} = \begin{bmatrix} 1 & 0 & 0 \\ 0 & c_{23} & s_{23} \\ 0 & -s_{23} & c_{23} \end{bmatrix} \begin{bmatrix} c_{13} & 0 & s_{13}e^{-i\delta} \\ 0 & 1 & 0 \\ -s_{13}e^{i\delta} & 0 & c_{13} \end{bmatrix} \begin{bmatrix} c_{12} & s_{12} & 0 \\ -s_{12} & c_{12} & 0 \\ 0 & 0 & 1 \end{bmatrix}. \tag{2.14}$$

The parameters of the CKM matrix can be determined experimentally, and the diagonal elements are close to 1 [4]. However, the small but non-vanishing off-diagonal elements lead to cross-generation decays, as illustrated in Fig. 2.3. The current world averages of the experimentally measured parameters of the SM are given in Table 2.1.

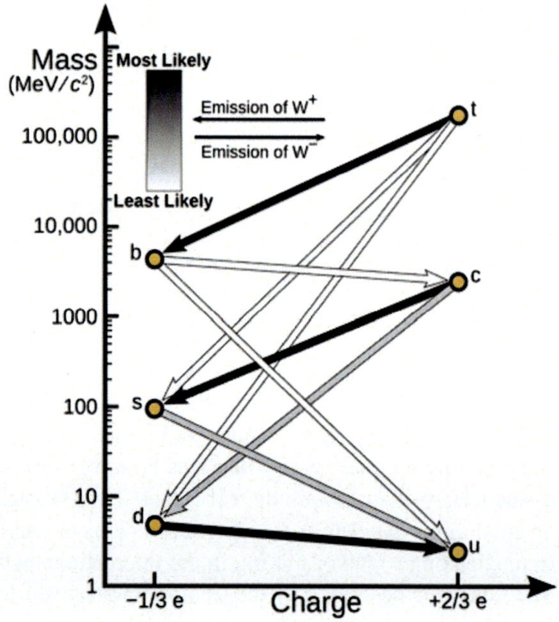

Fig. 2.3 The cross-generation quark decays in the SM. Moving from left (right) to right (left) implies the emission of a W^- (W^+) boson. The probability of a given decay route is illustrated by the darkness of the arrow. Image taken from Ref. [21]

2.2 The Stability of the Electroweak Vacuum

It is possible that the current minimum of the Higgs potential $V(H)$ is not stable, but that the field configuration might eventually change into another minimum. Since this would alter how the EW sector of the SM functions, it could have drastic cosmological implications. The issue arises from quantum loop corrections that could change $V(H)$, so that its derivative becomes negative at large values of the Higgs field, hence making the potential unbound from below [22].

Investigations of $V(H)$ at large values of the Higgs field require an improved form of the potential that is valid for a wide range of field values [22]. This can be obtained using renormalization group techniques [23, 24]. A particular requirement for stability is that the Higgs quartic coupling parameter λ should never become negative [22, 24, 25]. While the measured Higgs boson mass is close to the minimum value required for absolute stability in the context of the SM, the analysis of the potential implies a possibly vanishing λ around the Planck scale as illustrated in Fig. 2.4 (left) [26].

Since the coupling of the Higgs to other SM fields is strongest for the fields with the most massive particles, the most significant corrections to the parameters of $V(H)$ are caused by loops involving top quarks. Hence the evolution of λ depends on the values of α_S, m_t, and to a smaller extent on m_H [26–29]. The regions of instability, metastability and stability with respect to α_S and m_t are shown in Fig. 2.4 (right).

Table 2.1 The parameters of the SM and their experimentally measured values [4]. The top quark mass corresponds to the pole mass from cross section measurements. The strong coupling constant $\alpha_S(m_Z)$ is given instead of the $SU(3)$ coupling. Furthermore, the fine structure constant incorporates the elementary charge, and the couplings g_1 and g_2 can be expressed in terms of the Higgs vacuum expectation value and the weak boson masses

Parameter	Symbol	Value
Electron mass	m_e	$0.51099895000 \pm 3.1 \cdot 10^{-9}$ MeV
Muon mass	m_μ	$105.6583745 \pm 0.0000024$ MeV
Tau mass	m_τ	1776.86 ± 0.12 MeV
Up quark mass	m_u	$2.16^{+0.49}_{-0.26}$ MeV
Down quark mass	m_d	$4.67^{+0.48}_{-0.17}$ MeV
Strange quark mass	m_s	93^{+11}_{-5} MeV
Charm quark mass	m_c	1.27 ± 0.02 GeV
Bottom quark mass	m_b	$4.18^{+0.03}_{-0.02}$ GeV
Top quark mass	m_t	172.4 ± 0.7 GeV
W boson mass	m_W	80.379 ± 0.012 GeV
Z boson mass	m_Z	91.1876 ± 0.0021 GeV
Higgs boson mass	m_H	125.10 ± 0.14 GeV
Higgs vacuum expectation value	v	246.22 GeV
Fine structure constant	α	$(137.035999084 \pm 2.1 \cdot 10^{-8})^{-1}$
Strong coupling constant	$\alpha_S(m_Z)$	0.1179 ± 0.0010
sin of CKM mixing angle 12	s_{12}	0.22650 ± 0.00048
sin of CKM mixing angle 13	s_{13}	$0.00361^{+0.00011}_{-0.00009}$
sin of CKM mixing angle 23	s_{23}	$0.04053^{+0.00083}_{-0.00061}$
CP-violation phase	δ	$1.196^{+0.045}_{-0.043}$
QCD vacuum angle	θ_QCD	$\lesssim 10^{-10}$

The eventual settling of the question of vacuum stability will require precise experimental results for α_S, m_t and m_H. The phenomenological analyses presented in this thesis improve the precision of the values of α_S and m_t, which need to be determined experimentally.

2.3 Quantum Chromodynamics

In the SM, the strong interaction acts among quarks and is mediated by massless gluons. The strong interaction is described by the QFT of QCD, operating with *color charge*. QCD is a Yang-Mills theory [31] based on the $SU(N)$ gauge group with $N = 3$. There are eight gluon fields in QCD, corresponding to the eight generators of $SU(3)$. In the *fundamental representation*, the generators are typically written as $T^a \equiv \frac{1}{2}\lambda^a$. The matter fields transform in this representation,

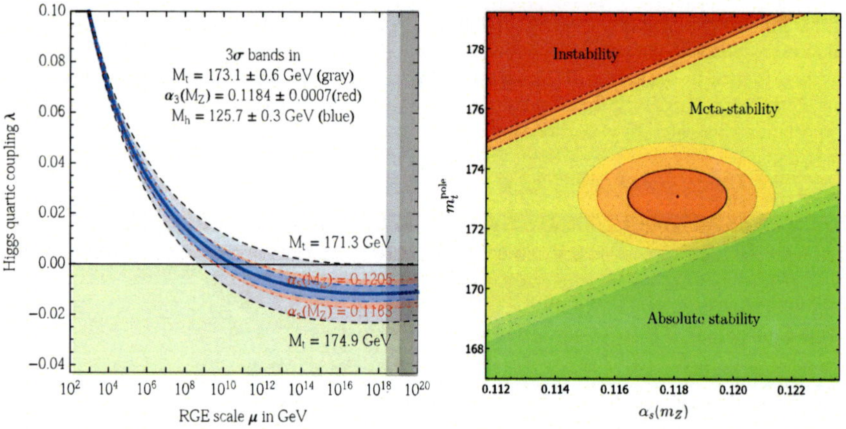

Fig. 2.4 *Left*: The evolution of the Higgs quartic coupling λ as a function of the renormalization group equation (RGE) scale μ. Image taken from Ref. [26]. *Right*: The regions of instability, metastability and stability with respect to the values of the strong coupling constant and the top quark pole mass. The dot corresponds to the values $m_t^{\text{pole}} = 173.1 \pm 0.6\,\text{GeV}$ and $\alpha_S(m_Z) = 0.1181 \pm 0.0011$, with the ellipses indicating the 68, 95 and 99% contours. Image taken from Ref. [30]

whereas the gauge fields transform in the *adjoint representation*, where the generators are given by $(T_A^a)^{bc} = -if^{abc}$, with the structure constants normalized by $f^{acd} f^{b}{}_{cd} = N\delta^{ab}$. The representations are characterized in a basis-independent way by the *Casimir operators*. In particular, the quadratic Casimir operator for a representation R is defined by $C_R \mathbb{1} \equiv T_R^a T_{Ra}$. For the fundamental representation, it reads $C_F = (N^2 - 1)/(2N)$, and $C_A = N$ for the adjoint representation. For QCD, they are then given by $C_F = 4/3$ and $C_A = 3$. Furthermore, the generators can be chosen so that $\text{Tr}[T_R^a T_R^b] = T_R \delta^{ab}$, where T_R is the representation's *index*. For the fundamental (adjoint) representation, the index reads $T_F = 1/2$ $(T_A = N)$ [1].

Writing the field strength tensor product in Eq. (2.5) open and including the mass terms explicitly, the QCD Lagrangian reads

$$\begin{aligned}
\mathscr{L}_{\text{QCD}} = & -\frac{1}{4}(\partial_\mu G_\nu^a - \partial_\nu G_\mu^a)(\partial^\mu G_a^\nu - \partial^\nu G_a^\mu) \\
& - \frac{g_3}{2} f_a{}^{bc}(\partial_\mu G_\nu^a - \partial_\nu G_\mu^a) G_b^\mu G_c^\nu - \frac{g_3^2}{4} f_a{}^{bc} f^a{}_{de} G_b^\mu G_c^\nu G_\mu^d G_\nu^e \\
& + \sum_{h \in \{R,L\}} \left(\overline{q}_h^i \gamma^\mu \partial_\mu \overline{q}_{hi} + \frac{g_3}{2} \lambda_a \overline{q}_h^i \gamma^\mu G_\mu^a q_{hi} \right) \\
& - m_i \left(\overline{q}_L^i q_{Ri} + \overline{q}_R^i q_{Li} \right) \\
& + \frac{1}{2\xi}(\partial_\mu G_a^\mu)(\partial^\nu G_\nu^a) + (\partial_\mu \overline{\chi}^a)\partial^\mu \chi^a + g_3 f^{abc}(\partial_\mu \overline{\chi}_a) G_b^\mu \chi_c,
\end{aligned} \tag{2.15}$$

where the q are color triplets and the sums over i account for the quark flavors [1, 32]. In addition to the terms following directly from Eq. (2.5) after the EWSSB, the last line of Eq. (2.15) contains a gauge fixing term dependent on the gauge parameter ξ. Choosing the gauge is necessary for defining the gluon propagator, and e.g. $\xi = 1$ in the Feynman-'t Hooft gauge and $\xi = 0$ in the Lorenz gauge [1]. Without fixing the gauge, the Green's functions of the elementary gauge-variant field operators vanish and the spontaneous breaking of a local symmetry is impossible in a symmetrical gauge theory without gauge fixing [33, 34]. The further additional terms on the last line of Eq. (2.15) introduce the *ghost* field χ, which is a complex scalar following Fermi statistics [35]. The ghosts do not correspond to physical particles, but are a strictly virtual concept for canceling non-physical degrees of freedom for the gluon [36].

The reach of perturbative quantum chromodynamics (pQCD) is approached at approximately the radius of the proton. This is related to the fundamental properties of QCD, which are *confinement* at large distances and *asymptotic freedom* at short distances. These properties arise from the underlying symmetry of QCD in the context of renormalization, as described in the following.

2.4 Perturbation Theory and Renormalization

The equations of motion (EOM) for a physical system are obtained by minimizing the dimensionless action $S = \int d^4x\mathscr{L}$. All terms in $\mathscr{L}$ must therefore have a total mass dimension of four to cancel the dimension 1/4 of d^4x. Masses, energy scales and derivatives are of dimension 1, whereas fermionic (bosonic) field operators have the dimension 3/2 (1) in 4-dimensional spacetime. Furthermore, the Lagrangian may contain dimensionless factors. In particular, the interaction terms involving three or more field operators contain the coupling constant of the respective interaction [1].

In perturbation theory, observables are computed by expanding them in terms of the coupling constant α_S in a region where it is small, and truncating the series at a chosen power of α_S; the lowest possible power yields leading order (LO) results, the second lowest gives next-to-leading order (NLO) results and so on. The amplitude of a scattering process can then be computed using the Feynman rules derived from the Lagrangian. However, taking a computation beyond LO increases the number of vertices in the graphs and allows for virtual loops and additional radiated particles, as illustrated for the case of QCD in Fig. 2.5. For instance at NNLO, the cross section of a process then takes the form

$$\sigma = \sigma^{(0)} + \frac{\alpha_S}{\pi}\sigma^{(1)} + \left(\frac{\alpha_S}{\pi}\right)^2 \sigma^{(2)} \tag{2.16}$$

where $\sigma^{(0)}$ is the LO result and the NLO term $\sigma^{(1)}$ is typically divided into virtual and real contributions as $\sigma^{(1)}_{\text{Virtual}} + \sigma^{(1)}_{\text{Real}}$. Analogously, the NNLO term $\sigma^{(2)}$ is divided into double-virtual, real-virtual or double-real contributions depending on if the NNLO

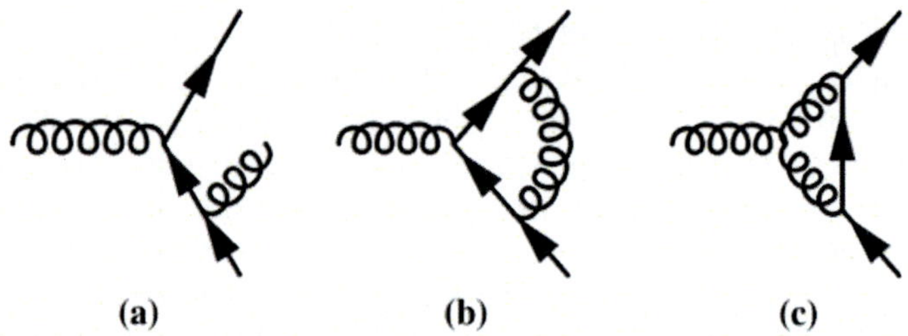

Fig. 2.5 Gluon splitting to a quark-antiquark pair at NLO. The convention of reading the graphs from left to right is chosen, and the arrows pointing forwards (backwards) correspond to quarks (antiquarks), whose interactions are mediated by the curly gluon lines. Graph **a** corresponds to real emission. The loops in **b** and **c** correspond to virtual corrections to the quark-gluon interaction vertex

graphs contain zero, one or two real emissions, respectively. Furthermore, the real emission cases where the radiation is connected to either the initial or final state particles are referred to as initial state radiation (ISR) or final state radiation (FSR), respectively.

The graphs beyond LO are not necessarily finite, but can diverge when high-momentum particles emit soft radiation, or a high-momentum particle is radiated collinear to a low-momentum particle, or when integrating over loop momenta. Cases where high particle momenta lead to singularities are called *ultraviolet* (UV) divergences, whereas *infrared* (IR) divergences arise from low momentum particles. According to the Kinoshita-Lee-Nauenberg theorem [37, 38], the IR divergences cancel in unitary theories when summing over all possible initial and final states in a finite energy window. *Renormalizable theories* are then defined as those in which all UV divergences can be absorbed into a finite number of parameters, such as couplings and masses [1]. According to the Bogoliubov-Parasiuk-Hepp-Zimmermann theorem, all the divergences of a renormalized QFT are cancelled by counterterms corresponding to superficially divergent one-particle irreducible (1PI) diagrams [39–41]. The process of renormalization however makes perturbative computations dependent on a choice of renormalization scale, and replaces the quantities used for absorbing the divergences by experimentally measured values, so they can no longer be predicted from first principles.

The diverging fields and parameters in a Lagrangian are to be considered *bare* quantities. For example, the bare fermion mass m^{bare} is divided into a finite renormalized mass m, and a factor Z_m that is to be regularized, via $m^{\text{bare}} = mZ_m$. Similarly, the field operators for quarks, gluons and ghosts are rewritten in the renormalization of QCD as $q^{\text{bare}} \equiv q\sqrt{Z_2}$, $G_\mu^{\text{bare}} \equiv G_\mu\sqrt{Z_3}$, and $\chi^{\text{bare}} \equiv \chi\sqrt{Z_\chi}$, where Z_2 is for the renormalization of the quark field, Z_3 for the gluons and Z_χ for the ghosts. Since the coupling appears in four types of vertices, as depicted in Fig. 2.6 [1], there are formally four Z factors associated with it: Z_1 for the quark-gluon coupling, $Z_{1\chi}$ for the ghost-gluon coupling, as well as Z_{G^3} and Z_{G^4} for the trilinear and quartic gluon self-couplings, respectively.

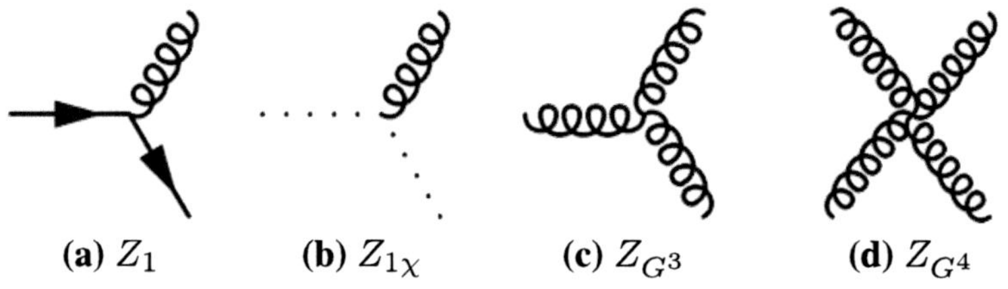

Fig. 2.6 The renormalization factors associated with the QCD coupling in quark-gluon interaction vertices (**a**), the ghost-gluon vertex (**b**) and the gluon self-coupling vertices (**c** and **d**)

Writing the bare quantities in a Lagrangian in terms of the renormalized ones and the Z factors then results in terms that have the form of the original Lagrangian, but written using the renormalized quantities, and new terms incorporating the Z factors, leading to Feynman rules for counterterms [1]. For instance, the Dirac Lagrangian for bare quarks can be rewritten as

$$\begin{aligned}\mathscr{L}_{\text{Dirac}} &= i\overline{q}^{\text{bare}}\gamma^\mu\partial_\mu q^{\text{bare}} - m^{\text{bare}}\overline{q}^{\text{bare}}q^{\text{bare}} \\ &= iZ_2\overline{q}\gamma^\mu\partial_\mu q - Z_m Z_2 m\overline{q}q \\ &= \underbrace{i\overline{q}\gamma^\mu\partial_\mu q - m\overline{q}q}_{\text{Original form of } \mathscr{L}_{\text{Dirac}}} + \underbrace{i(Z_2-1)\overline{q}\gamma^\mu\partial_\mu q - (Z_m Z_2 - 1)m\overline{q}q}_{\Longrightarrow \text{ Counterterms}}\,. \end{aligned} \tag{2.17}$$

This has introduced new terms that have the form of quark self-interactions, and provide the counterterms, so that a finite value for the NLO quark 2-point function can be calculated from

(2.18)

The Z coefficients are chosen in the course of the calculation, so that they cancel the divergence arising from integrating over all loop momenta. However, the requirement that the counterterms cancel the divergences does not determine their finite parts, which are affected by the choice of renormalization prescription. E.g. in the minimal subtraction (MS) scheme, the counterterms have no finite parts, only the terms necessary for canceling all divergences.

2.4.1 Regularization and the Renormalization Scale

The process of *regularization* is to perform the divergent integrals in a way that isolates the divergent contributions into poles depending on a regulator ε, which can be taken to zero once the poles are identified and the counterterms are chosen suitably to cancel them. One method for regulating the UV divergences in a gauge-invariant way is *dimensional regularization* (DR) [42, 43]. It makes a replacement of the form

$$g^2 \int \frac{d^4k}{(2\pi)^4} \rightarrow g^2 \mu_r^{4-d} \int \frac{d^dk}{(2\pi)^d} \tag{2.19}$$

for the integrals over loop momentum k. The 4-dimensional space-time is replaced by a d-dimensional one, since an integral divergent in 4 dimensions may be calculable in $d = 4 - 2\varepsilon$ dimensions, and results for non-integer ε are obtained by analytic continuation [1]. However, this also changes the dimensions of the bosonic (ϕ) and fermionic (ψ) fields to $d/2 - 1$ and $(d-1)/2$, respectively, and e.g. a fermion-boson interaction of the form $g\overline{\psi}\psi\phi$ would imply the coupling constant g to have dimension $2 - d/2$. The couplings of a renormalizable theory must nonetheless be dimensionless [1]. In DR this is achieved by introducing the t'Hooft scale μ_r of dimension 1, which is identified as the *renormalization scale*. Removing the regulator by taking the $\varepsilon \rightarrow 0$ limit then leads to expansions, where the divergent parts depending on ε are separated into poles with a strength of at most $1/\varepsilon^L$ for normal UV divergences at L loops [33].

Computations are commonly carried out in the modified minimal subtraction ($\overline{\text{MS}}$) scheme [44], in which μ_r is replaced by $\mu_r \exp(\varepsilon\gamma_E)(4\pi)^{-\varepsilon}$ with $\gamma_E \approx 0.577$ the Euler-Mascheroni constant. Although the scheme is unphysical, it has the advantages of being suitable also for massless theories, and preserving complicated symmetries, with the exception of those that cannot in general be preserved by quantization [45].

Besides DR, the $\overline{\text{MS}}$ prescription applies to other regulators as well. A common alternative is the *Pauli-Villars regularization* [46], where each particle of mass m receives a ghost particle with mass Λ. The ghost particle has either a wrong-sign kinetic term in the Lagrangian, or interchanged bosonic and fermionic statistics. This way, the ghost loop contributions have the same form but different sign as the particle loops' UV divergences, and the process will introduce terms dependent on $\log(m^2/\Lambda^2)$ to the result. Yet another option is *lattice regularization*, where continuous space-time is replaced by a discrete lattice and the Yang-Mills action is written in terms of Wilson loops [47].

If the sum of the operators' mass dimensions in a term in the Lagrangian would exceed 4, it would require a factor with a negative mass dimension. A theory with such terms is *non-renormalizable*. Despite the name, such theories can provide predictions also beyond LO, and are valid for a limited spectrum of energies. One instance of a non-renormalizable theory is discussed in Sect. 3.2.

2.4.2 Renormalization Group Equations

The computation of physical observables in fixed-order perturbation theory requires a truncation of the perturbative series. Therefore, the renormalization scale enters the calculations of physical observables in perturbation theory. The choice of scale will then introduce a source of uncertainty in theoretical predictions, which can be viewed as being due to missing higher order corrections.

However, if a scale μ is not a physical parameter itself, it could be demanded that a physical observable X is independent of it. This is expressed as $\frac{\partial}{\partial\mu}X^{\text{bare}} = 0$. Likewise, theories with an UV cutoff Λ should be required to produce the same predictions at energy scales much below Λ, regardless of the precise selection of Λ. Such differential equations are the basic idea of renormalization group equations (RGEs). The RGE for a quantity X generally has the form

$$\mu\frac{dX}{d\mu} = \gamma_X X \tag{2.20}$$

where γ_X is the *anomalous dimension*. For instance, the μ_r independence of the bare mass yields

$$0 = \mu_r\frac{dm^{\text{bare}}}{d\mu_r} \Leftrightarrow 0 = Z_m m\left(\frac{\mu_r}{m}\frac{dm}{d\mu_r} + \frac{\mu_r}{Z_m}\frac{dZ_m}{d\mu_r}\right) \implies \gamma_m = -\frac{\mu_r}{Z_m}\frac{dZ_m}{d\mu_r} \tag{2.21}$$

where the mass anomalous dimension γ_m has been identified [1].

Solving an RGE allows for evolving the theory from one scale selection to another so that the couplings in the theory change accordingly, while the observable stays invariant. Strictly, the renormalization group is not a mathematical group, but refers to a trajectory in the space of possible theories that result in the same physical observables [1].

2.5 The Strong Coupling Constant

Although called a constant for conventional reasons, the coupling α_S of the strong interactions acquires a scale dependence in the renormalization of QCD [48]. The strong coupling constant is conventionally renormalized by introducing the factor Z_1 for the quark-gluon interaction vertex[1] and accounting for the two $\sqrt{Z_2}$ and one $\sqrt{Z_3}$ factors from the renormalization of the fields, so that the bare coupling is written as

$$g^{\text{bare}} = \mu_r^{\frac{4-d}{2}} g\frac{Z_1}{Z_2\sqrt{Z_3}}. \tag{2.22}$$

[1] Renormalizing via the gluon self-interaction vertex yields the same results due to gauge invariance. In general, the charge is universal in Yang-Mills theories [1]. In principle, the four Z-factors in Fig. 2.6 provide the counterterms for the vertices in QCD, but computations are simplified by the Slavnov-Taylor identities [49, 50] generalizing the Ward-Takahashi identities of QED [51, 52] to non-Abelian theories and providing relations between different Z factors.

At NLO, the graphs contributing to Z_1 are those given in Fig. 2.5b and c and the vertex counterterm, while the factor Z_2 is obtained along with Z_m from the quark self-energy graphs given in Eq. (2.18). The NLO graphs contributing to the factor Z_3 are

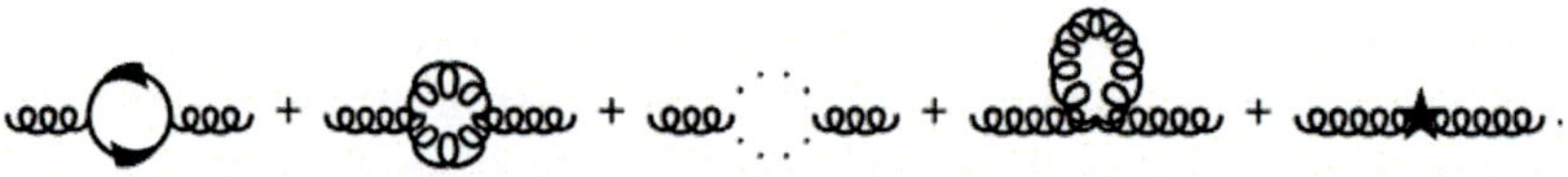

(2.23)

In $d = 4 - 2\varepsilon$ dimensions, the factors are given by [1, 33]

$$\begin{aligned} Z_1 &= 1 - \frac{\alpha_S S_\varepsilon}{4\pi\varepsilon}\left[\frac{3-\xi}{4}C_A + \xi C_F\right] + \mathcal{O}(\alpha_S^2), \\ Z_2 &= 1 - \frac{\alpha_S S_\varepsilon}{4\pi\varepsilon}\xi C_F + \mathcal{O}(\alpha_S^2), \\ Z_3 &= 1 - \frac{\alpha_S S_\varepsilon}{4\pi\varepsilon}\left[\left(\frac{\xi}{2} - \frac{13}{6}\right)C_A + \frac{4}{3}T_F n_f\right] + \mathcal{O}(\alpha_S^2), \end{aligned} \tag{2.24}$$

where

$$S(\varepsilon) = \frac{(4\pi)^\varepsilon}{\Gamma(1-\varepsilon)} \approx 1 + \varepsilon\left[\log(4\pi) - \gamma_E\right]. \tag{2.25}$$

Demanding $dg^{\text{bare}}/d\mu_r = 0$ leads to an RGE for the *running* QCD coupling. This RGE is related to the *QCD beta function* [53], which is given by

$$\mu^2 \frac{da_S}{d\mu^2} = -\sum_{n=0} \beta_n \left(a_S(\mu)\right)^{n+2}, \tag{2.26}$$

with

$$a_S \equiv \alpha_S/\pi, \tag{2.27}$$

and the coefficients β_n up to 3 loops in the $\overline{\text{MS}}$ scheme read [54–62]

$$\begin{aligned} \beta_0 &= \frac{1}{4}\left(\frac{11}{3}C_A - \frac{2}{3}n_f\right), \\ \beta_1 &= \frac{1}{16}\left(\frac{34}{3}C_A^2 - 2C_F n_f - \frac{10}{3}C_A n_f\right), \\ \beta_2 &= \frac{1}{64}\left(\frac{2857}{54}C_A^3 + C_F^2 n_f + \frac{205}{18}C_F C_A n_f \right. \\ &\quad \left. + \frac{1415}{54}C_A^2 n_f + \frac{11}{9}C_F n_f^2 + \frac{79}{54}C_A n_f^2\right), \end{aligned} \tag{2.28}$$

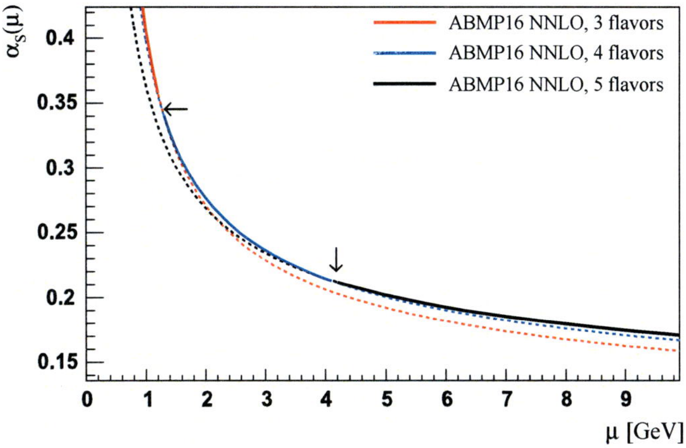

Fig. 2.7 The α_S running in the ABMP16 PDF at NNLO [68]. The solid lines represent the cases with 3, 4 and 5 active flavors in the regions where α_S is evaluated at $\mu < m_c$, $m_c < \mu < m_b$ and $\mu < m_b$, respectively. The flavor thresholds are indicated by arrows. The dashed lines extrapolate the α_S with different numbers of active flavors, illustrating their deviance in the various regions

where the number of active flavors is denoted by n_f. The factors of $1/4^{n+1}$ in each β_n are due to following the conventions of Ref. [63]. The coefficients β_n have also been computed at four [64, 65] and five loops [66].

Note that crossing a quark flavor threshold, i.e. evaluating α_S at a quark mass m_Q, requires comparing a theory with n_f flavors to one with $n_f + 1$ flavors. In general, this gives rise to a non-trivial matching of $\alpha_S^{(n_f)}(m_Q)$ to $\alpha_S^{(n_f+1)}(m_Q)$, although $\alpha_S^{(n_f)}(m_Q) = \alpha_S^{(n_f+1)}(m_Q)$ up to NLO in the $\overline{\text{MS}}$ scheme [67]. The effect of n_f on the running of $\alpha_S(\mu)$ is illustrated in Fig. 2.7, showing the regions where different numbers of active flavors would be relevant. An overview of the experimental measurements of the running of α_S is shown in Fig. 2.8.

At 1-loop level, the β function has the solution

$$a_S(\mu) = \frac{1}{\beta_0 \log(\mu^2/\Lambda_{\text{QCD}}^2)}, \tag{2.29}$$

where Λ_{QCD} is the location of the *Landau pole* in QCD, and perturbation theory becomes inapplicable for $\mu < \Lambda_{\text{QCD}}$ since $a_S(\mu)$ diverges. For $n_f = 5$, the pole is approximately at 213 MeV [1]. The divergence of a_S at small scales leads to confinement: an attempt to separate two quarks by more than 1 fm, the typical length scale of a hadron, will result in the energy between the quarks to become sufficient for producing new hadrons. Hence, quarks are never observed as free particles, but always constitute composite particles. Nonetheless, this process of *hadronization* cannot be predicted from first principles in QCD, but must be modeled phenomenologically. In contrast, the negative sign in Eq. (2.26) leads to the $a_S(\mu)$

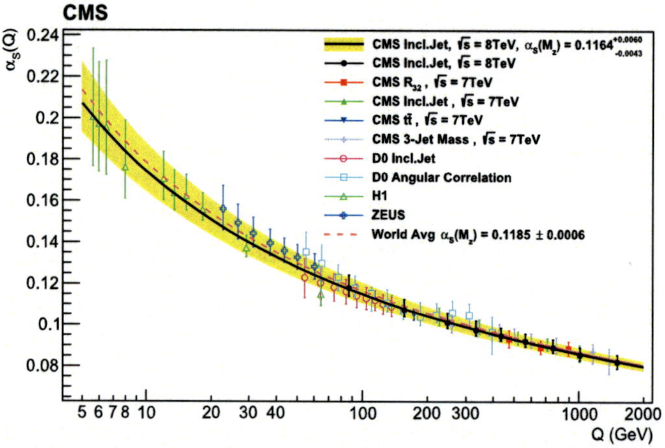

Fig. 2.8 The strong coupling constant α_S as a function of $Q = p_T$, determined at 2 loops in Ref. [69] (black line). The band represents its total uncertainty. Values of $\alpha_S(Q)$ extracted by the CMS Collaboration in several ranges of Q are shown, along with the H1, ZEUS, and D0 Collaborations' results from HERA and Tevatron. Image taken from Ref. [70]

getting progressively weaker at large μ. This asymptotic freedom makes it plausible to involve quarks in the initial and final states of perturbative calculations in the high-energy regime.

At higher orders, α_S can be estimated by perturbative expansions around the LO solution. Furthermore, the value of α_S at a scale μ' can be estimated by expanding it about a reference scale μ as [1]

$$
\begin{aligned}
a_S(\mu') =& a_S(\mu) + (a_S(\mu))^2 \beta_0 \log\left(\frac{\mu^2}{\mu'^2}\right) \\
&+ (a_S(\mu))^3 \left[\beta_1 \log\left(\frac{\mu^2}{\mu'^2}\right) + \beta_0^2 \log^2\left(\frac{\mu^2}{\mu'^2}\right)\right] + \mathcal{O}((a_S(\mu))^4). \quad (2.30)
\end{aligned}
$$

2.6 The Renormalization of Quark Masses

Since quarks are not observed as free particles, their masses are quantities defined formally via renormalization. This can be done using various schemes: while the pole mass is the pQCD equivalent of the on-shell mass of a free particle, there are various other schemes with unique advantages and disadvantages in terms of the interpretation of experimental results. Some definitions introduce a scale dependence, causing also the quark masses to run, analogous to the strong coupling constant. The running phenomenon has been demonstrated for the mass of the b quark in Ref. [71], using data from experiments at the LEP [72–75], SLC [76], and HERA [77] colliders. Also the running of the c quark mass has been demonstrated experimentally using HERA data [78]. Furthermore, the running of the mass of the top quark has been determined recently by the CMS Collaboration [79].

2.6.1 *The Pole Mass Scheme*

The pole, or *on-shell*, scheme is defined via the renormalization of the 2-point Green's function, computed from the graphs [1]

$iG_2(\not{p}) =$ 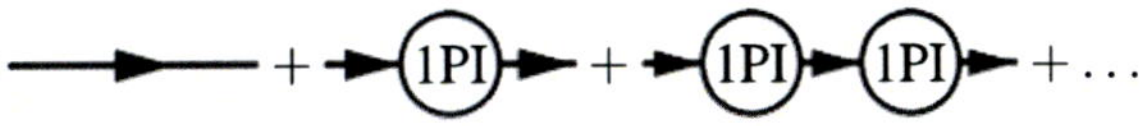

where $\not{p} = \gamma^\mu p_\mu$ and 1PI stands for the sum of one-particle irreducible loop graphs

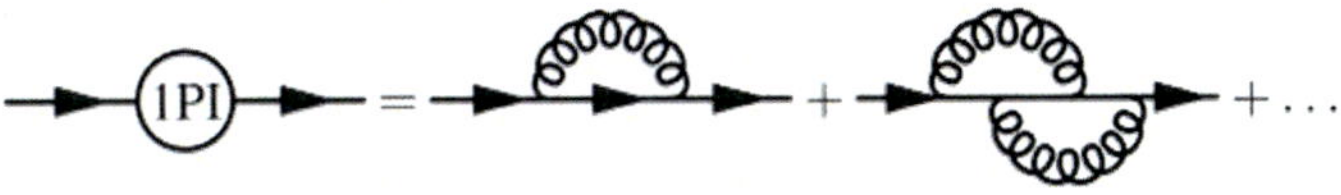

so that

$$iG_2(\not{p}) = \frac{i}{\not{p} - m + \Sigma(\not{p})} \tag{2.31}$$

where $\Sigma(\not{p})$ is defined as the sum of all 1PI graphs. After renormalizing the Green's function, all 1PI diagrams are summed into it, and the location of the pole can be identified as the physical mass m_P for a free particle. For the renormalized Σ_R and m_R, this implies the conditions [1]

$$\begin{cases} \Sigma(m_P) = m_R - m_P \\ \frac{d}{d\not{p}}\Sigma(\not{p})\Big|_{\not{p}=m_P} = 0 \end{cases} \tag{2.32}$$

However, the pole mass suffers from a linear IR sensitivity known as the *renormalon ambiguity*, which is of $\mathcal{O}(\Lambda_{\text{QCD}})$ [80–82]. It is therefore attractive to consider the so-called short-distance mass schemes, in particular for physical observables dominated by short distances. These do not suffer from the renormalon ambiguity, and lead to a scale-dependence resulting in the phenomenon of running for the mass. In this thesis, the $\overline{\text{MS}}$ and MSR short-distance mass schemes are investigated.

2.6.2 The $\overline{\text{MS}}$ Mass Scheme

In computations using the $\overline{\text{MS}}$ scheme, the mass renormalization scale is often set equal to μ_r. However, also a general mass scale μ_m can be used, in which case the pole and $\overline{\text{MS}}$ masses are related by

$$m_t^{\text{pole}} = \overline{m}_t(\mu_m)\left(1 + \sum_{n=1} d_n^{\overline{\text{MS}}}(\mu_m)(a_S(\mu_m))^n\right). \tag{2.33}$$

When the decoupling $\alpha_S^{(6)} \to \alpha_S^{(5)}$ is performed at $\overline{m}(\mu_m)$ the decoupling coefficients d_n relevant for computations up to NNLO in QCD with $SU(3)$ are given by

$$\begin{aligned} d_1^{\overline{\text{MS}}}(\mu_m) &= \frac{4}{3} + L, \\ d_2^{\overline{\text{MS}}}(\mu_m) &= \frac{307}{32} + 2\zeta_2 + \frac{2}{3}\zeta_2 \log 2 - \frac{1}{6}\zeta_3 + \frac{509}{72}L + \frac{47}{24}L^2 \\ &\quad - n_f\left(\frac{71}{144} + \frac{1}{3}\zeta_2 + \frac{13}{36}L + \frac{1}{12}L^2\right), \end{aligned} \tag{2.34}$$

where $L \equiv \log((\mu_m/\overline{m}(\mu_m))^2)$, n_f is the number of active light quarks and ζ_n is the value of the Riemann zeta function at n. The running of the $\overline{\text{MS}}$ mass is described by the RGE

$$\mu^2 \frac{d\overline{m}}{d\mu^2} = -\overline{m} \sum_{i=0} \gamma_i^m (a_S(\mu))^{i+1} \tag{2.35}$$

with the anomalous dimensions

$$\begin{aligned} \gamma_0^m &= 1, \\ \gamma_1^m &= \frac{1}{16}\left(\frac{202}{3} - n_f \frac{20}{9}\right), \\ \gamma_2^m &= \frac{1}{64}\left(1249 - n_f\left[\frac{2216}{27} + \frac{160}{3}\zeta_3\right] - n_f^2 \frac{140}{81}\right), \end{aligned} \tag{2.36}$$

for QCD with $SU(3)$ [63, 83–87]. The RGE in Eq. (2.35) has the solution

$$\overline{m}(\mu_1) = \overline{m}(\mu_0) \exp\left\{-2 \sum_{i=0} \int_{\mu_0}^{\mu_1} d\mu \gamma_i^m \frac{(a_S(\mu))^{i+1}}{\mu}\right\}. \tag{2.37}$$

Solving Eq. (2.37) yields the $\overline{\text{MS}}$ mass at a scale μ_1 via a proper evolution procedure, when the mass is known at a given reference scale μ_0. Typically the $\overline{\text{MS}}$ mass is reported at the scale of the quark mass itself as $\overline{m}(\overline{m})$.

2.6.3 The MSR Mass Scheme

The MSR scheme was first introduced in Ref. [88], then discussed conceptually in relation to the Monte Carlo masses in Ref. [89], and a detailed discussion is given in Ref. [90]. The pole and MSR masses are related by

$$m^{\text{pole}} = m^{\text{MSR}} + R \sum_{n=1}^{\infty} d_n^{\text{MSR}} (a_S(R))^n, \tag{2.38}$$

where the d_n^{MSR} are decoupling coefficients and the scale R serves to make the MSR mass approach the definition of the pole mass when $R \to 0$, and to the $\overline{\text{MS}}$ mass $\overline{m}(\overline{m})$ at $R = \overline{m}(\overline{m})$. A feature of evolving the scale R in the MSR scheme is that for the difference of MSR masses at two scales, the linear renormalization scale dependence provides an all-orders resummation of the terms in the asymptotic series associated with the pole mass renormalon ambiguity. Since the pole mass renormalon problem is related to self-energy corrections at the scale $\Lambda_{\text{QCD}} < R$, the MSR mass can be interpreted as having a mass renormalization constant containing the on-shell self-energy corrections only for scales larger than R. While all self-energy corrections for quantum fluctuations up to scales m are absorbed into the pole mass, $m^{\text{MSR}}(R)$ only absorbs those between R and m [89, 90].

In the context of this thesis, the MSR scheme is expected to be applied for the top quark mass primarily in the region $R < \overline{m}_t(\overline{m}_t)$. Therefore, it is appropriate to change the scheme from one including the top quark UV effects with $5 + 1$ active flavors, to a scheme with only $n_f = 5$ active flavors. This can be done by either rewriting $a_S^{(n_f+1)}$ in terms of $a_S^{(n_f)}$, which results in the *practical MSR* (MSRp) mass definition, or by integrating out the virtual loop corrections of the top quark, resulting in the *natural MSR* (MSRn) mass definition [90].

For MSRp, the values of the decoupling coefficients d_n^{MSR} in Eq. (2.38) are[2]

$$\begin{aligned} d_1^{\text{MSRp}} &= 4/3 \\ d_2^{\text{MSRp}} &= 13.443375 - 1.04136875 n_f \\ d_3^{\text{MSRp}} &= 190.390625 - 26.65515625 n_f + 0.652690625 n_f^2, \end{aligned} \tag{2.39}$$

whereas for MSRn they read

$$\begin{aligned} d_1^{\text{MSRn}} &= 4/3 \\ d_2^{\text{MSRn}} &= 13.3398125 - 1.04136875 n_f \\ d_3^{\text{MSRn}} &= 188.671875 - 26.67734375 n_f + 0.652690625 n_f^2, \end{aligned} \tag{2.40}$$

[2] The coefficients d_n^{MSR} in this section correspond to the decoupling factors a_n given in Ref. [90], but divided by 4^n to account for a difference in the conventions of Chap. 5 and Ref. [90].

which are in agreement with those of the $\overline{\text{MS}}$ scheme when the number of heavy quarks is set to zero. In the case of MSRp, the first two coefficients agree with the $\overline{\text{MS}}$ coefficients when the number of heavy quarks is set to one. Here, only the top quark is treated as massive and the five active light flavors are taken to be massless. The matching of the $\overline{\text{MS}}$ mass with $n_f + 1$ active flavors to MSRn with n_f flavors is given at $R = \overline{m}_t$ by [90]

$$m_t^{\text{MSRn}}(\overline{m}_t) = \overline{m}_t(\overline{m}_t)\times \Big[1 + 0.103567\left(a_S^{(n_f)}(\overline{m}_t)\right)^2 + (1.71953 + 0.2225 n_f)\left(a_S^{(n_f)}(\overline{m}_t)\right)^3 + (1.34375 - 0.435898 n_f + 0.0171875 n_f^2)\left(a_S^{(n_f)}(\overline{m}_t)\right)^4\Big]. \tag{2.41}$$

Conversely, $\overline{m}_t(\overline{m}_t)$ is obtained from a value of $m_t^{\text{MSR}}(m_t^{\text{MSR}})$ via [90]

$$\overline{m}_t(\overline{m}_t) = m_t^{\text{MSR}}\left(m_t^{\text{MSRn}}\right)\times \Big[1 - 0.103567\left(a_S\left(m_t^{\text{MSRn}}\right)\right)^2 - (1.58141 + 0.02225 n_f)\left(a_S\left(m_t^{\text{MSRn}}\right)\right)^3 + (1.39453 + 0.403711 n_f - 0.0171875 n_f^2)\left(a_S\left(m_t^{\text{MSRn}}\right)\right)^4\Big]. \tag{2.42}$$

For MSRp, the matching relation is $m_t^{\text{MSRp}}(\overline{m}_t) = \overline{m}_t(\overline{m}_t)$. The MSR mass at a scale R is obtained by using the appropriate matching relation to take a given $\overline{\text{MS}}$ mass to MSRn or MSRp, and evolving the scale from $\overline{m}_t$ to R. The MSR mass evolution is governed by [90]

$$R\frac{d}{dR} m_t^{\text{MSR}}(R) = -R\sum_n \gamma_n^R (a_S(R))^{n+1}. \tag{2.43}$$

The anomalous dimensions needed for converting the mass at $\mathcal{O}(\alpha_S^3)$ read [90]

$$\begin{aligned} \gamma_0^R &= d_1^{\text{MSR}}, \\ \gamma_1^R &= d_2^{\text{MSR}} - 2\beta_0 d_1^{\text{MSR}}, \\ \gamma_2^R &= d_3^{\text{MSR}} - 4\beta_0 d_2^{\text{MSR}} - 2\beta_1 d_1^{\text{MSR}}, \end{aligned} \tag{2.44}$$

where β_n are the coefficients of the QCD beta-function given in Sect. 2.5. The MSR mass at R is then $m_t^{\text{MSR}}(R) = m_t^{\text{MSR}}(\overline{m}_t) + \Delta m$, where the difference Δm is obtained via the solution of Eq. (2.43) as

$$m_t^{\text{MSR}}(\overline{m}_t) - m_t^{\text{MSR}}(R) = -\sum_{n=0}^{2} \gamma_n^R \int_R^{\overline{m}_t} dR' \left(a_S(R')\right)^{n+1} + \mathcal{O}(a_S^4) \equiv \Delta m. \quad (2.45)$$

In the context of this thesis, the first study of the behavior of the scale R is performed. This serves to assess and improve the stability of $t\bar{t}$ production cross sections as a function of the scales μ_f, μ_r and R, which is particularly important for improving the precision in the extraction of the top quark mass from experimental data.

2.7 Phenomenology of Jet and Top Quark Physics

The quarks and gluons of QCD, altogether referred to as *partons*, are not observed as free particles, but constitute baryons and mesons as hadronic bound states. To satisfy the Pauli exclusion principle, the partons carry a degree of freedom called the color charge [91], which cancels among the partons in a hadron. Except for the top quark, the partons produced in high energy collisions emit further partons, and eventually the partons form color-neutral hadrons. This results in collimated sprays of hadrons called *jets*, which preserve the properties of the original parton - its flavor, momentum and direction.

Observations of jet production were of utmost importance to establishing QCD. The production of two jets in e^-e^+ collisions is consistent with the production of spin-1/2 quarks, which subsequently hadronize and form 2 jets, showcasing the confinement property of QCD [92]. The existence and spin of the gluon were revealed by the production of three-jet events via the Drell-Yan process $e^-e^+ \to \gamma^*/Z \to q\overline{q}g$, predicted in Ref. [93] and observed in Ref. [94]. Experimental proof for the non-Abelian nature of QCD was obtained from the production of 4 jets in e^+e^- collisions, as the process is sensitive to the trilinear gluon vertex color factor C_A [92]. Experimental measurements of the ratios C_A/C_F and T_F/C_F are in agreement with the theoretical values corresponding to $SU(3)$ [95].

The clustering of the produced hadrons into jets can however be done in multiple ways, and there are important properties that a suitable algorithm should account for. In principle, the boundaries of a jet are undefined, and contributions from the showering of a hard parton may be missed if the jet size is small. However, large jet size may in turn lead to contamination from additional particles not belonging to the jet. Furthermore, the algorithm should be *infrared-collinear safe*, ergo robust against additional emissions of soft or collinear gluons. In contemporary analyses, the clustering is commonly performed using the infrared-collinear safe *recombination algorithms*, which start by finding the minimum of the set of distances between particles i and j as

$$d_{ij} \equiv \min\left(k_{Ti}^{2p}, k_{Tj}^{2p}\right) \frac{\Delta y_{ij}^2 + \Delta\phi_{ij}^2}{\mathcal{R}^2}, \quad (2.46)$$

and between particle i and the beam B as $d_{iB} \equiv k_{\mathrm{T}i}^{2p}$. Here k_{T} is the transverse momentum, p is a parameter defining the employed jet algorithm, and $\mathcal{R}$ is the cone radius parameter describing the size of the jet, while $\Delta\phi_{ij}^2$ and Δy_{ij}^2 are the squared differences between the azimuthal angles ϕ and rapidities y of i and j, when

$$y = \frac{1}{2}\frac{E + p_z}{E - p_z}. \tag{2.47}$$

Here E is an object's energy and p_z the momentum along the beam. The algorithms proceed iteratively to check if $d_{ij} < d_{iB}$, in which case i and j are clustered into a jet by summing their momenta; otherwise i is labeled a jet. If all particles are not included in a jet, the set of distances is redefined and the process repeated.

There are three common recombination algorithms, corresponding to different choices of the parameter p in Eq. (2.46). The k_{T} algorithm [96], obtained with the setting $p = 1$, assumes that the momenta of the particles inside a jet are similar. The irregular shapes of the jets are caused by clustering the soft particles first, and the algorithm is sensitive to particles produced in the vicinity of the jet. The choice $p = 0$ corresponds to the *Cambridge-Aachen* algorithm [97], which is also sensitive to the surrounding effects, but conserves jet substructure. Finally, the *anti-k_{T}* algorithm [98] with $p = -1$ clusters the hard particles first and results in jet shapes that are stable even with nearby activity. The jets obtained from the same event by the three different recombination algorithms are illustrated in Fig. 2.9.

In numerical computations, partonic scattering is calculated using perturbation theory, and the emission of soft and collinear partons is simulated using *parton shower* programs to approximate higher-order terms with a perturbative approach. Besides the partonic scatterings at large-momenta, and the subsequent parton branchings, every pp collision event gives rise to multiple parton-parton interactions leading to soft QCD processes. This is referred to as the *underlying event*. Such effects cannot be calculated perturbatively, but are simulated with phenomenological models tuned to experimental data [99]. Further, there are color correlations between the final state partons, with potentially significant effects to the kinematics [100]. This is addressed by *color reconnection*, for which different models have been developed [101, 102].

Fixed-order pQCD predictions for inclusive jet production cross sections are available up to NNLO, and are discussed in Sect. 6.6.1. However, the pQCD predictions contain neither hadronization nor multiparton interactions. These nonperturbative (NP) effects are corrected with a factor

$$\mathrm{NP} = \frac{\sigma^{\mathrm{MC}}(\mathrm{PS\ \&\ MPI\ \&\ HAD})}{\sigma^{\mathrm{MC}}(\mathrm{PS})}, \tag{2.48}$$

where PS stands for parton shower, HAD for hadronization, and MPI for multiparton interactions. At low p_{T}, the NP corrections are dominated by MPI, which increase the radiation in the jet cone by a constant offset; this is especially important for large-radius jets. Since the effects of perturbative radiation are partially considered in higher-order predictions, the PS simulation is included both in the numerator and

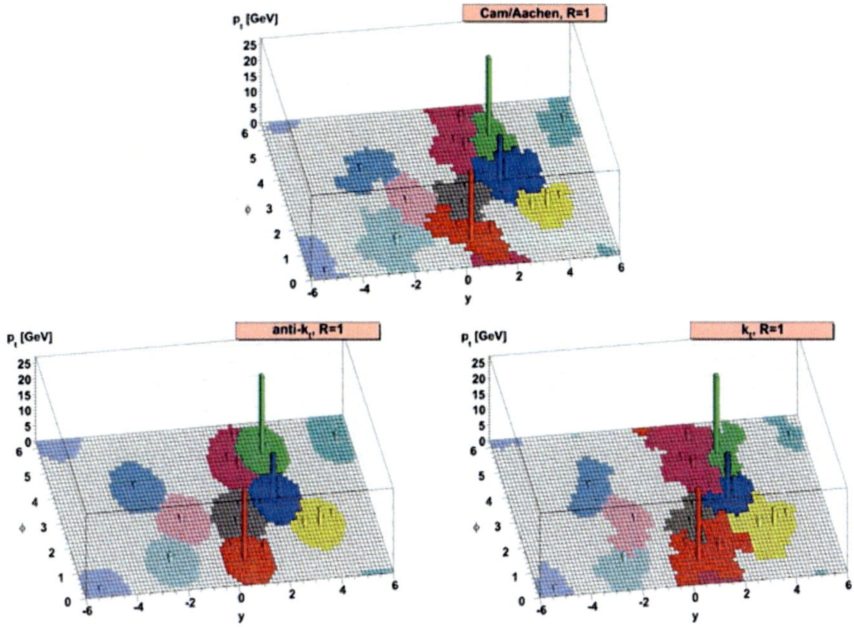

Fig. 2.9 The jets reconstructed in a sample event by using different jet algorithms. Images taken from Ref. [98]

denominator of Eq. (2.48). It is stronger for smaller $\mathcal{R}$, where out-of-cone radiation plays a larger role, which next-to-leading logarithmic (NLL) corrections can account for. Furthermore, the virtual exchange of massive weak gauge bosons W and Z requires the QCD predictions for jet production to be corrected for EW effects. In the high-p_{T} region, these grow up to 11% at high p_{T} and low y [103]. The EW and NP corrections for inclusive jet production cross sections in pp collisions at $\sqrt{s} = 13$ TeV are illustrated in Fig. 2.10.

In addition to jet production, the phenomenological investigations in this thesis address the production of top quarks and antiquarks, with the aim of a global interpretation of the LHC data. The top quark was first observed in 1995 by the CDF [104] and D0 [105] Collaborations at the Fermilab Tevatron in $\mathrm{p\bar{p}}$ collisions at $\sqrt{s} = 1.8$ TeV. With a pole mass of 172.4 ± 0.7 GeV obtained from cross section measurements [4], it is the most massive elementary particle in the SM. The decay time of the top quark is significantly below the time scale of hadronization, making it possible to study the properties of an unconfined quark. For instance, the polarization and charge of the top quark have been measured in Refs. [106, 107].

In pp collisions, top quarks are mainly produced via $\mathrm{t\bar{t}}$ pair production, and about 90% of the $\mathrm{t\bar{t}}$ pairs at the LHC are produced via the gluon-gluon fusion process [109] with the LO graphs

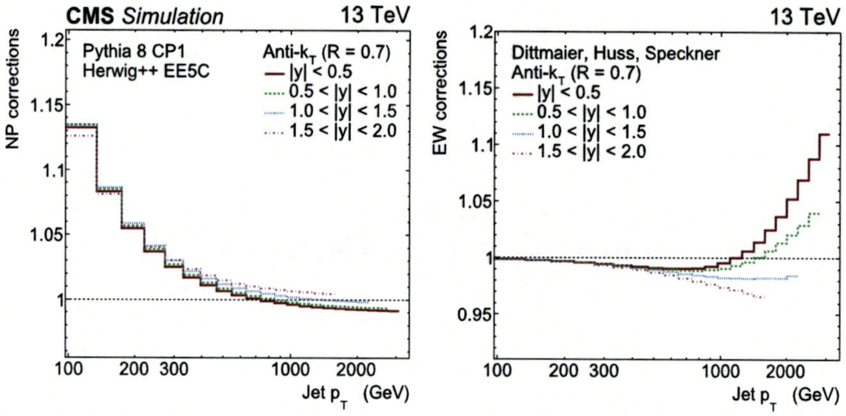

Fig. 2.10 *Left*: the values for NP corrections for inclusive jet cross sections, with each curve corresponding to a rapidity bin [108]. *Right*: the EW corrections for inclusive jet cross sections, as reported in Ref. [103], for jets clustered using the anti-k_T algorithm with $\mathcal{R} = 0.7$

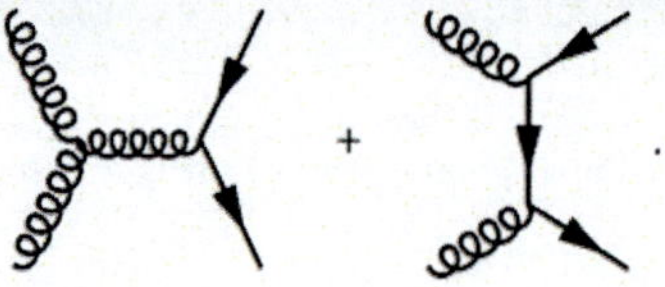

(2.49)

Details of the computation of fixed-order predictions for $t\bar{t}$ production cross sections are given in Chap. 5. The top quark decays via the charged-current weak interaction, and the branching ratio for the main decay channel $t \to Wb$ is obtained from the components of the CKM matrix as

$$B(t \to Wb) = \frac{|V_{tb}|^2}{|V_{tb}|^2 + |V_{ts}|^2 + |V_{td}|^2}. \tag{2.50}$$

The $t\bar{t}$ processes are classified into different channels, based on the W boson decays. In the fully hadronic, or all-jets channel, both W bosons decay hadronically. In the dileptonic channel, both W bosons decay into a lepton and a corresponding neutrino. In the semileptonic channel, one of the W bosons decays leptonically and the other one hadronically. The respective branching ratios are illustrated in Fig. 2.11.

Contributions to the production cross section of single top quarks or antiquarks are generally divided into three types: the Mandelstam t and s channels, and associated production resulting in a W + t final state. These are illustrated at LO in Fig. 2.12.

At the LHC, single top production is dominated by the t-channel, which is also responsible for approximately 20% of all t-quark production at the LHC [111]. However, also the contribution of the associated W + t production plays an important role at LHC energies [112]. The contributions of the different channels are illustrated in Fig. 2.13.

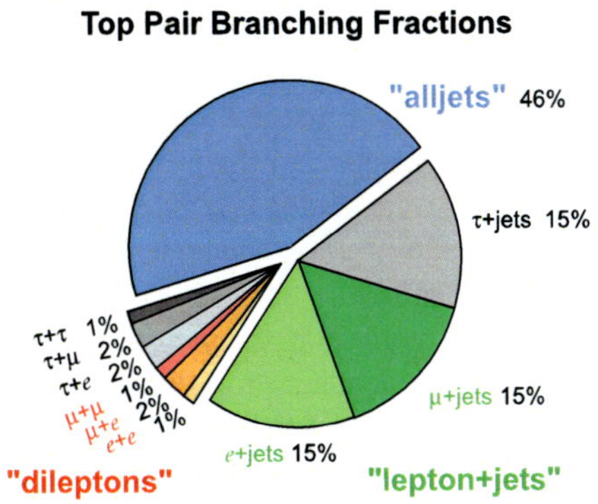

Fig. 2.11 The branching ratios of the various decay channels of $t\bar{t}$. The label *lepton+jets* stands for the semileptonic channels, and *alljets* for the fully hadronic channel. Image taken from [110]

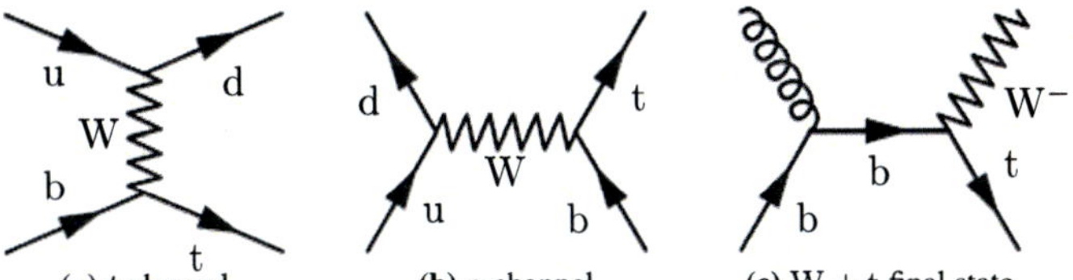

(a) t-channel **(b)** s-channel **(c)** $\mathrm{W} + \mathrm{t}$ final state

Fig. 2.12 Examples of LO diagrams where the single quark line in the final state can correspond to a t quark, while the antiquark has a different flavor due to the flavor mixing in the electroweak interactions mediated by the W boson

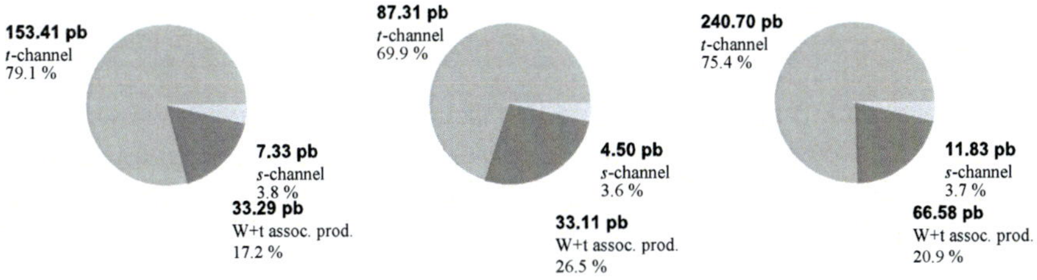

(a) t-production **(b)** $\bar{t}$-production **(c)** t and $\bar{t}$

Fig. 2.13 The contributions of the t and s channel and $\mathrm{W} + \mathrm{t}$ associated production to the single t-quark production cross section (**a**), to single $\bar{t}$ production (**b**) and when accounting for both t and $\bar{t}$ (**c**). All cross sections are computed at NLO, for pp collisions at $\sqrt{s} = 13\,\mathrm{TeV}$, and using the HATHOR default CKM matrix and EW parameter values. The results are obtained in the $\overline{\mathrm{MS}}$ scheme with $\overline{m}_t = 160.68\,\mathrm{GeV}$ and using the ABMP16 [68] PDF set at NLO

The measurements of m_t based on the reconstruction of the invariant mass of the decay products are known as *direct measurements*, and can reach extreme precision. However, direct measurements rely on distributions derived from Monte Carlo (MC) simulations. This makes them sensitive to the properties of the utilized generator and parton shower, since the simulation does not only contain the fixed-order matrix elements but also ISR, FSR, hadronization and underlying event interactions, which are modeled by parton showers and heuristic models [113]. Particularly, m_t is affected by the presence of an infrared cutoff Q_0 in the parton shower, implying that the MC mass could be considered a Q_0-dependent short-distance mass and would hence not correspond to the pole mass [114]. Nonetheless, connecting the MC mass to well-established short-distance schemes suffers from the insufficient accuracy of leading-logarithmic (LL) pQCD-based parton showers to fix the scheme, as the differences between schemes only appear at NLL accuracy [115]. The MC mass is then merely understood as the value of the mass parameter m_t^{MC} in the MC simulation, giving the best description of experimental data.

Nevertheless, the MC mass can be calibrated into a QFT mass scheme. In Ref. [113], this is done via a generic approach for measuring an observable ξ sensitive to m_t in some renormalization scheme, without assumptions on m_t^{MC} or its relation to m_t. A simultaneous likelihood fit of m_t^{MC} and ξ is performed, comparing an observed distribution in data to the prediction from MC. The value of m_t is then determined in the given renormalization scheme by comparing data to theoretical predictions for $\xi(m_t)$, and m_t^{MC} is calibrated by quantifying the difference $\Delta_m = m_t - m_t^{\mathrm{MC}}$. In Ref. [116], a calibration is made by fitting the predictions of a given MC generator to hadron level QCD computations, for observables that are related closely to the distributions entering the experimental analysis. It has been suggested that there can be a difference of $\mathcal{O}(1\,\mathrm{GeV})$ between m_t^{MC} and m_t^{pole}, and that the value of m_t^{MC} corresponds to that of the MSR scheme, with the scale R set close to the MC shower cut-off scale [88, 89, 116]. In Ref. [117], a relation between m_t^{MC} and $m_t^{\mathrm{MSR}}(R = 1\,\mathrm{GeV})$ is found by fitting NLL predictions to the jet mass distribution obtained from particle-level Monte Carlo samples, resulting in

$$m_t^{\mathrm{MSR}}(R = 1\,\mathrm{GeV}) = 172.42 \pm 0.1\,\mathrm{GeV}, \tag{2.51}$$

when POWHEG is interfaced with the PYTHIA 8 generator, and

$$m_t^{\mathrm{MSR}}(R = 1\,\mathrm{GeV}) = 172.27 \pm 0.09\,\mathrm{GeV}, \tag{2.52}$$

when POWHEG is interfaced with HERWIG 7. Note that in Ref. [117], the scales are set to $\mu_f = \mu_r = \sqrt{m_t^2 + p_T^2}$, with m_t and p_T the mass and transverse momentum of the top quark. The difference between the m_t^{MC} in the POWHEG + PYTHIA 8 simulation and the MSR mass at the scale $R = 1\,\mathrm{GeV}$ is then determined to be [117]

$$m_t^{\mathrm{MC}} - m_t^{\mathrm{MSR}}(1\,\mathrm{GeV}) = 80^{+350}_{-400}\,\mathrm{MeV}. \tag{2.53}$$

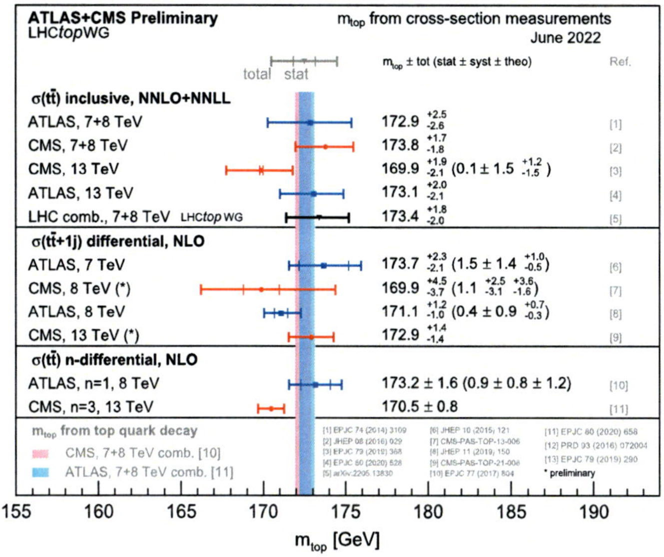

Fig. 2.14 A summary of the measurements of the top quark mass from $t\bar{t}$ production observables by ATLAS and CMS [123]

As an alternative to the direct measurements, a theoretically well-defined mass, such as the pole or a short-distance mass, can be determined by comparing measured observables sensitive to m_t with fixed-order predictions for e.g. the $t\bar{t}$ cross section [118–122]. A summary of the values of the top quark mass measured using $t\bar{t}$ production observables at the LHC is shown in Fig. 2.14.

At the LHC, the QFT top quark mass is mostly extracted by using measurements of inclusive or differential cross sections of $t\bar{t}$ production. However, in pp collisions these cross sections are not only driven by the value of m_t, but also by the description of the proton structure and the value of α_S. In Ref. [124], the normalized triple-differential $t\bar{t}$ production cross section measured by the CMS Collaboration in pp collisions at $\sqrt{s} =$ 13 TeV [125] and the HERA DIS data [126] are used in a simultaneous extraction of $m_t^{\text{MSR}}(R = 3\,\text{GeV})$, $\alpha_S(m_Z)$ and the parton distributions. The theoretical predictions for the $t\bar{t}$ production cross section are computed in the pole mass scheme with the scale choice $\mu_r = \mu_f = m_t^{\text{pole}}$, and translated into the MSR scheme via [124]

$$\sigma^{\text{MSR}}\left(m_t^{\text{MSR}}(R)\right) = \sigma^{\text{pole}}\left(m_t^{\text{pole}}\right)\Big|_{m_t^{\text{pole}}=m_t^{\text{MSR}}(R)} + \left(m_t^{\text{MSR}}(R) - m_t^{\text{pole}}\right)\left(\frac{d\sigma^0}{dm}\right)\Big|_{m=m_t^{\text{MSR}}(R)}, \tag{2.54}$$

where σ^0 is the Born contribution to the cross section, and the difference between the pole and MSR masses is computed up to $\mathcal{O}(\alpha_S^3)$. This results in the values $m_t^{\text{MSR}}(R =$

3 GeV) $= 169.6^{+0.8}_{-1.1}$ GeV and $\alpha_S(m_Z) = 0.1132^{+0.0023}_{-0.0018}$, where the uncertainties are computed by adding the individual uncertainties reported in Ref. [124] in quadrature.

The structure of the proton, in particular the gluon distribution, plays an important role in jet and top quark physics alike. This is detailed in the following.

2.8 The Structure of the Proton in the LHC Era

With a mass of $m_\mathrm{p} = 938.3$ MeV [4] the proton is the lightest baryon, and generally considered stable due to baryon number conservation in the SM [1]. Further, there is no experimental evidence of the energetically allowed decays of the proton to lighter particles. The current lower bound for the proton lifetime is $3.6 \cdot 10^{29}$ years [127], while decay modes e.g. $p \rightarrow e + \pi^0$ result in a 90% CL limit of $1.6 \cdot 10^{34}$ years [128].

The quantum numbers of the proton arise from one d and two u *valence quarks*, whose concept was suggested in [129–131]. Besides these, the quark-parton model (QPM) [132, 133], developed in the late 1960s contains also *sea quarks* with no overall flavor, emitted and absorbed continuously within the proton. The predictions of the QPM agreed with the deep inelastic scattering (DIS) measurements carried out at the Stanford linear accelerator center (SLAC) [134]. However, the model could only explain approximately half of the protons total momentum [135]. In contrast to the QPM with non-interacting quarks, the quarks in QCD interact via gluon exchange.

Due to confinement, the structure of hadrons cannot be calculated from first principles only, using pQCD. A connection between the observable cross sections from hadron collisions and pQCD predictions is given by the *factorization theorem* [136]. The master formula for computing cross sections in hadron collisions factorizes the perturbative partonic cross section $\hat{\sigma}_{ij}$ from the structure of the colliding protons, which is encoded in the parton distribution functions (PDFs) f_i, f_j, as

$$\sigma = \int dx_1 \int dx_2 \sum_{ij} \mathrm{f}_i(x_1\,, \mu_f)\ \mathrm{f}_j(x_2\,, \mu_f)\hat{\sigma}_{ij}\left(x_1, x_2, \alpha_S(\mu), \mu_r, \mu_f\right). \quad (2.55)$$

Here the sums over i and j account for the types and flavors of partons. The PDF f_i is related to the probability of finding the parton i, within a fraction range Δx of the proton momentum, inside the proton as resolved at factorization scale μ_f. The PDFs are taken to be universal functions of the momentum fraction x, and the factorization scale defining the resolution below which the physics can be absorbed into the nonperturbative PDFs. Some features of factorization are illustrated in Fig. 2.15.

The relevant RGEs for evolving the PDFs are the Dokshitzer-Gribov-Lipatov-Altarelli-Parisi (DGLAP) equations [138–143]. For n_f flavors, the $2n_f - 1$ scalar non-singlet evolution equations read

$$\mu\frac{d}{d\mu}\mathrm{f}_{q_i}(x, \mu) = \sum_j \int_x^1 \frac{d\xi}{\xi} P_{q_i q_i}(x/\xi)\mathrm{f}_{q_i}(\xi, \mu), \quad i \in [\mathrm{v}, \pm] \quad (2.56)$$

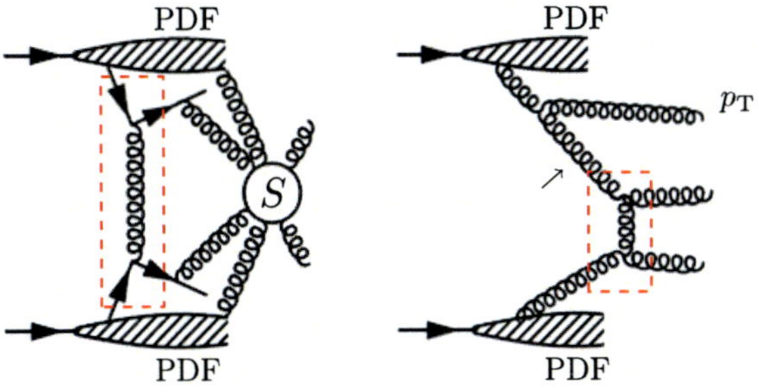

Fig. 2.15 *Left*: an example of factorization for jet production in a pp collision. The hard process is surrounded by a rectangle drawn in dashes, while the shaded blobs correspond to the PDFs. The effects of the soft gluons connected to the S blob cancel in factorization for inclusive cross sections [137]. *Right*: integrating over p_T leads to a collinear divergence related to the factorization scale μ_f, caused by the gluon line indicated by an arrow [137]

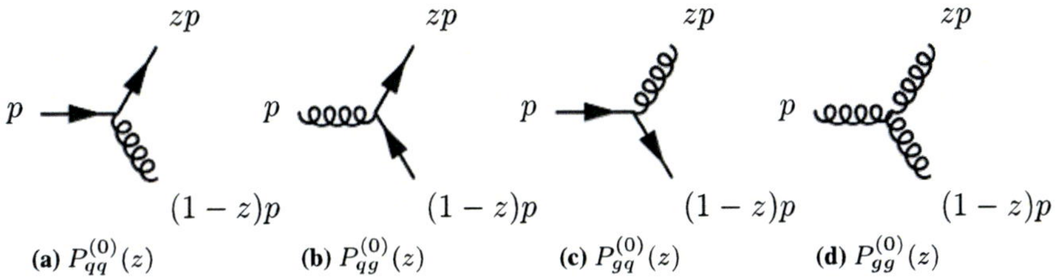

Fig. 2.16 A representation of LO splitting functions, with p an off-shell momentum

for the valence distributions q_v and the non-singlet quark flavor asymmetries $q_\pm$, while the 2×2 singlet evolution reads

$$\mu \frac{d}{d\mu} \begin{bmatrix} \mathrm{f}_{q_i}(x, \mu) \\ \mathrm{f}_g(x, \mu) \end{bmatrix} = \sum_j \int_x^1 \frac{d\xi}{\xi} \begin{bmatrix} P_{q_i q_j}(x/\xi) & P_{q_i g}(x/\xi) \\ P_{g q_j}(x/\xi) & P_{gg}(x/\xi) \end{bmatrix} \begin{bmatrix} \mathrm{f}_{q_j}(\xi, \mu) \\ \mathrm{f}_g(\xi, \mu) \end{bmatrix}, \qquad (2.57)$$

for the singlet quark distributions, for which i and j run over the quark flavors, and for the gluon distribution denoted by the index g [1, 144]. The splitting functions $P_{ab}(z)$ are expanded in terms of α_S/π, and at LO they have an interpretation as the probability of obtaining a parton a, with a fraction z of the parent parton momentum, from parton b, and into another quark or gluon (as deducible from knowing a and b) with momentum fraction $1-z$ [139]. The splitting functions are obtained perturbatively, and the graphs for the LO splitting functions are shown in Fig. 2.16.

At LO, due to charge conjugation invariance and flavor symmetry, the splitting functions are independent of the quark flavor, and have the same forms for quarks and antiquarks [145], so that $P_{q_i q_j} = P_{\overline{q}_i \overline{q}_j}$, $P_{q_i \overline{q}_j} = P_{\overline{q}_i q_j}$, $P_{q_i g} = P_{\overline{q}_i g}$ and $P_{g q_i} = P_{g \overline{q}_i}$. However, these relations are violated at higher orders [142]. Furthermore, interpreting the splitting functions as probabilities implies that they are positive definite for $z < 1$, and they satisfy the *sum rules*

$$\begin{cases} \int_0^1 dz P_{qq}^{(0)}(z) = 0, \\ \int_0^1 dz z \left(P_{qq}^{(0)}(z) + P_{gq}^{(0)}(z)\right) = 0, \\ \int_0^1 dz z \left(2n_f P_{qg}^{(0)}(z) + P_{gg}^{(0)}(z)\right) = 0, \end{cases} \tag{2.58}$$

corresponding to the conservation of quark number and momentum [145]. Here the quark flavor indices have been suppressed. Similarly, the PDFs obey a set of sum rules [145, 146]. The conservation of the number of up- and down quarks in the proton and the conservation of momentum correspond to

$$\begin{cases} \int_0^1 dx[\mathrm{u}(x) - \bar{\mathrm{u}}(x)] = 2, \\ \int_0^1 dx[\mathrm{d}(x) - \bar{\mathrm{d}}(x)] = 1, \\ \sum_j \int_0^1 dx \mathrm{f}_j(x) x = 1. \end{cases} \tag{2.59}$$

The DGLAP equations evolve μ_f while keeping x fixed. On the other hand, evolving x and keeping μ_f fixed is referred to as Balitsky-Fadin-Kuraev-Lipatov evolution, which however is closer to the perturbative region and only considered to work for low x and high (but finite) μ_f [147]. Since the full x-dependence of the PDFs cannot be predicted from first principles in pQCD, it needs to be extracted from fits to experimentally measured cross sections. The PDFs are typically parametrized in the form a polynomial at a chosen starting scale Q_0, and then evolved numerically by the DGLAP equations to the kinematic region of the measurements. The parameters in the PDF parametrization are then iterated until the fit converges.

The general shape of the gluon and quark singlet distributions is such that they have high values at low x and decrease rapidly at high x, while the valence quark distributions have a peak at medium x values and go to zero as $x \to 0$ and $x \to 1$. The characteristic form of DGLAP evolution is such that parton distributions tend to move towards lower x values as the PDFs are evolved to higher μ_f. This can be interpreted so that high-x partons radiate other partons with lower x values, and the parent partons drop to smaller x in the process. This can be pictured as an increase in the amount of partons in the proton, as its structure is probed at higher scales [146].

The PDF sets provided by different phenomenology groups, such as ABMP, CTEQ, HERAPDF and NNPDF, are referred to as *global PDFs*. Each group uses different selections of data, theory assumptions and parametrizations for the functional form of the PDFs at the starting scale. As an example, the valence quark, sea and gluon distributions of the HERAPDF2.0 PDF are shown in Fig. 2.17.

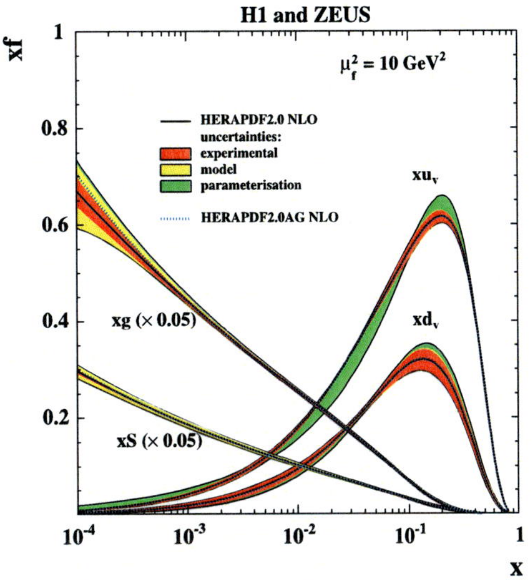

Fig. 2.17 The valence quark (xu_v, xd_v), sea (xS) and gluon (xg) PDFs in the HERAPDF2.0 at NLO [148]. The PDFs are evaluated at $\mu_f^2 = 10\,\text{GeV}^2$. The gluon and the sea distributions are scaled down by a factor of 20. Image taken from Ref. [148]

2.8.1 Constraining the PDFs with Experimental Data

The main constraints on PDFs are set by experimental fits to the DIS data measured at HERA. The kinematics of the e^-p scattering events, from which the structure of the proton is measured, are commonly described in terms of a standard set of variables. The *virtuality* $Q^2 \equiv -q^2 = -(k-k')^2$ indicates the resolution capabilities of the process, so that more of the proton structure can be resolved with rising Q. The *inelasticity* $Y \equiv P_\mu q^\mu/(P_\nu k^\nu)$ gives the fraction of momentum transferred from the scattering lepton to the hadronic system in the proton rest frame, and the fraction of the momentum of the proton participating in the reaction is given by $x = Q^2/(2P_\mu q^\mu)$ [145]. The contributions to the $e^-p \to e^-X$ process are divided into *neutral current* (NC) and *charged current* (CC) processes. As illustrated in Fig. 2.18, and the cross sections are given in terms of three structure functions F_1, F_2 and F_3. The neutral current contribution is given by

$$\frac{d^2\sigma^{\text{NC}}}{dxdQ^2} = \frac{4\pi\alpha^2}{xQ^4}\big[xY^2F_1^{\text{NC}}(x,Q^2) + (1-Y)F_2^{\text{NC}}(x,Q^2) + Y(1-Y/2)xF_3^{\text{NC}}(x,Q^2)\big] \tag{2.60}$$

where α is the EM coupling constant [145]. The charged current contribution is given by

$$\frac{d^2\sigma^{\text{CC}}}{dxdQ^2} = \frac{(1-\lambda_e)\pi\alpha^2}{8\sin^4\theta_W(Q^2+m_W^2)^2}\sum_{i,j}\big[|V_{u_id_j}|^2\text{u}_i(x) + |V_{u_jd_i}|^2\text{d}_i(x)(1-Y)^2\big], \tag{2.61}$$

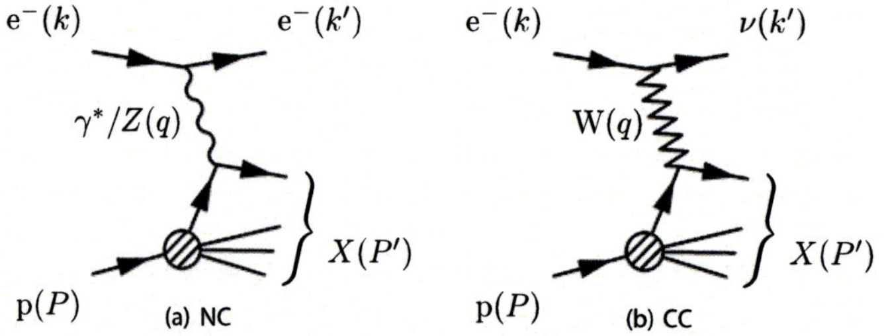

Fig. 2.18 Neutral current (**a**) and charged current (**b**) processes in e^-p collisions. The hadronic final state is denoted by X, and the four-momenta are indicated in brackets

where λ_e is the electron helicity, u_i (d_i) refer to up (down) type quarks, $V_{u_i d_j}$ are the CKM matrix elements and the weak mixing angle can be given in terms of the $U(1)$ and $SU(2)$ couplings as $\theta_W = g_1/\sqrt{g_1^2 + g_2^2}$ [145].

The measurements of various processes in proton-proton collisions at the LHC provide complementary information in different kinematic regions, and enhance the precision of the PDFs. Instead of the momentum fraction x as in e^-p, the kinematics of pp collisions depend on two momentum fractions x_1 and x_2, corresponding to the two incoming protons. The production of jets in hadron collisions is sensitive to the gluon PDF at high-x, and top quark pair-production cross sections provide complementary information on the gluon PDF in the medium and high-x region. The production of EW bosons probes different bilinear combinations of light quarks, shedding light on the valence quark distributions and the flavor separation in the sea. The charge asymmetry in $W^\pm$ production improves the constraints on valence quark distributions. The shape of the Z rapidity distribution, and the W/Z ratio are sensitive to the strange quark PDF. Furthermore, the high-mass Drell-Yan process is sensitive to the photon PDF. The production of EW bosons associated with c quarks provides information on the strange quark content of the proton. The sensitivities of the cross sections measured at the LHC to different PDFs are summarized in Table 2.2, and an illustration of the kinematic reach of different measurements is illustrated in Fig. 2.19.

A proper understanding of the structure of the proton is of paramount importance in the interpretation of hadron collider measurements. As a historical example, in an analysis of Tevatron run I data at $\sqrt{s} = 1.8$ TeV [149], the experimental data were seen to be significantly above the theoretical predictions at high energy scales. However, instead of being a smoking gun for new physics, later studies of CDF run II data at 1.96 TeV revealed the measurement to be compatible with the predictions, after nonperturbative parton-to-hadron corrections were included [150].

There is however also another risk of bias regarding the extraction and use of the PDFs. The effects of many models beyond the Standard Model (BSM) are expected to manifest in the high-p_T and low rapidity region. However, the PDFs are extracted from experimental data in the same kinematic region, assuming the validity of the SM. Hence, there is a risk of absorbing the effects of new physics in the PDF fit.

Table 2.2 The sensitivity of LHC processes to different PDFs

Process	Sensitivity
Drell-Yan	Flavor decomposition of the sea, u_v, d_v and photon PDF
W+c, W+jets	Strange PDF
Vector boson + jets	Medium-x gluon PDF
Jets	High-x gluon and quark PDFs
Photon	Medium-x gluon PDF
Top pair	Medium- and high-x gluon PDF
Single top	High-x u/d ratio

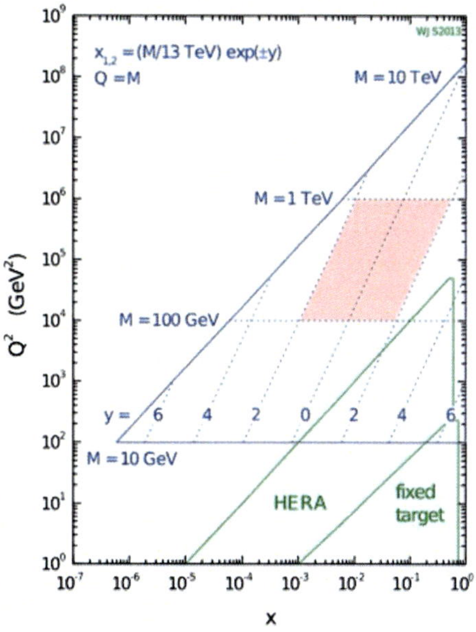

Fig. 2.19 The kinematic regions covered by HERA, fixed target and the CMS 13 TeV inclusive jet data. The latter are represented by the shaded area. Image after Ref. [151]

To assess this issue, the QCD analysis in Chap. 7 follows an *unbiased strategy* in a search for contact interactions (CI), where the PDFs and SM parameters are extracted simultaneously with the CI Wilson coefficients while using effective theory corrections for the jet cross section prediction.

References

1. Schwartz MD (2014) Quantum field theory and the standard model. ISBN: 978-1-107-03473-0, 978- 1-107-03473-0. Cambridge University Press, Cambridge
2. Cottingham WN, Greenwood DA (2007) An introduction to the standard model of particle physics, 2nd ed. Cambridge University Press, Cambridge. https://doi.org/10.1017/CBO9780511791406
3. Aitchison I, Hey T (2004) Gauge theories in particle physics: a practical introduction, vol 2. Non-Abelian Gauge Theories: QCD and the electroweak theory
4. Particle Data Group, Zyla PA et al (2020) Review of particle physics. Prog Theor Exp Phys 2020:083C01
5. Peccei RD, Quinn HR (1977) CP conservation in the presence of pseudoparticles. Phys Rev Lett 38:1440–1443
6. Peccei RD, Quinn HR (1977) Constraints imposed by CP conservation in the presence of pseu doparticles. Phys Rev D 16:1791–1797
7. Wilczek F (1978) Problem of strong P and T invariance in the presence of instantons. Phys Rev Lett 40:279–282
8. Weinberg S (1978) A new light boson? Phys Rev Lett 40:223–226
9. Asano Y et al (1981) Search for a rare decay mode $K^+ \to \pi^+$ neutrino anti-neutrino and axion. Phys Lett B 107:159. (Cence RJ, Ma E, Roberts A (eds))
10. Kim JE (1979) Weak-interaction singlet and strong CP invariance. Phys Rev Lett 43:103–107
11. Shifman MA, Vainshtein AI, Zakharov VI (1980) Can confinement ensure natural CP invariance of strong interactions? Nucl Phys B 166:493–506
12. Dine M, Fischler W, Srednicki M (1981) A simple solution to the strong CP problem with a harm less axion. Phys Lett B 104:199–202. ISSN: 0370-2693
13. Zhitnitsky AR (1980) On possible suppression of the axion hadron interactions. (In Russian). Sov J Nucl Phys 31:260
14. The free media repository WC (2022) Standard model of elementary particles. Accessed: 07 Jan 2022. https://commons.wikimedia.org/w/index.php?curid=4286964
15. The free media repository WC (2007) Standard model of elementary particles. Accessed 13 Feb 2022. https://commons.wikimedia.org/wiki/File:Elementary_particle_interactions.svg
16. Higgs PW (1964) Broken symmetries, massless particles and gauge fields. Phys Lett 12
17. Englert F, Brout R (1964) Broken symmetry and the mass of the gauge vector mesons. Phys Rev Lett 13
18. Aad G et al (2012) Observation of a new particle in the search for the standard model Higgs boson with the ATLAS detector at the LHC. Phys Lett B 716:1–29
19. Chatrchyan S et al (2012) Observation of a new boson at a mass of 125 GeV with the CMS experiment at the LHC. Phys Lett B 716:30–61
20. Neubauer M (2021) Higgs potential. Accessed 27 Dec 2021. https://msneubauer.github.io/projects/4_project/
21. The free media repository WC (2013) Weak decay. Accessed 07 Jan 2022. https://commons.wikimedia.org/wiki/File:Weak_Decay_(flipped).svg
22. Altarelli G, Isidori G (1994) Lower limit on the Higgs mass in the standard model: an update. Phys Lett B 337:141–144
23. Sher M (1989) Electroweak Higgs potentials and vacuum stability. Phys Rept 179:273–418
24. Ford C, Jones DRT, Stephenson PW, Einhorn MB (1993) The Effective potential and the renormalization group. Nucl Phys B 395:17–34
25. Cabibbo N, Maiani L, Parisi G, Petronzio R (1979) Bounds on the fermions and Higgs boson masses in grand unified theories. Nucl Phys B 158:295–305
26. Degrassi G et al (2012) Higgs mass and vacuum stability in the standard model at NNLO. JHEP 08:098
27. Alekhin S, Djouadi A, Moch S (2012) The top quark and Higgs boson masses and the stability of the electroweak vacuum. Phys Lett B 716:214–219

28. Bezrukov F, Kalmykov MY, Kniehl BA, Shaposhnikov M (2012) Higgs boson mass and new physics. JHEP 10:140 (Moortgat-Pick G (ed))
29. Bednyakov AV, Kniehl BA, Pikelner AF, Veretin OL (2015) Stability of the electroweak vacuum: gauge independence and advanced precision. Phys Rev Lett 115:201802
30. Andreassen A, Frost W, Schwartz MD (2018) Scale invariant instantons and the complete life time of the standard model. Phys Rev D 97:056006
31. Yang CN, Mills RL (1954) Conservation of isotopic spin and isotopic gauge invariance. Phys Rev 96:191–195
32. Buchalla G, Buras AJ, Lautenbacher ME (1996) Weak decays beyond leading logarithms. Rev Mod Phys 68:1125–1244
33. Collins J (2011) Foundations of perturbative QCD. Cambridge University Press, New York, NY. https://doi.org/10.1017/CBO9780511975592. https://cds.cern.ch/record/1350496
34. Elitzur S (1975) Impossibility of spontaneously breaking local symmetries. Phys Rev D 12:3978–3982
35. Faddeev L, Popov V (1967) Feynman diagrams for the Yang-Mills field. Phys Lett B 25:29–30. ISSN: 0370-2693
36. Rabbertz K (2017) Jet physics at the LHC. ISBN: 978-3-319-42113-1. https://doi.org/10.1007/978-3-319-42115-5
37. Kinoshita T (1962) Mass singularities of Feynman amplitudes. J Math Phys 3:650–677
38. Lee TD, Nauenberg M (1964) Degenerate systems and mass singularities. Phys Rev 133:B1549–B1562
39. Bogoliubov NN, Parasiuk OS (1957) On the multiplication of the causal function in the quantum theory of fields. Acta Math 97:227–266
40. Hepp K (1966) Proof of the Bogolyubov-Parasiuk theorem on renormalization. Commun Math Phys 2:301–326
41. Zimmermann W (1969) Convergence of Bogolyubov's method of renormalization in momentum space. Commun Math Phys 15:208–234
42. 't Hooft G, Veltman MJG (1972) Regularization and renormalization of gauge fields. Nucl Phys B 44:189–213
43. Bollini CG, Giambiagi JJ (1972) Dimensional renormalization: the number of dimensions as a regularizing parameter. Nuovo Cim B 12:20–26
44. Bardeen WA, Buras AJ, Duke DW, Muta T (1978) Deep-inelastic scattering beyond the leading order in asymptotically free gauge theories. Phys Rev D 18:3998–4017
45. Collins JC (1984) Renormalization: an introduction to renormalization, the renormalization group, and the operator-product expansion. Cambridge University Press, Cambridge https://doi.org/10.1017/CBO9780511622656. https://cds.cern.ch/record/105730
46. Pauli W, Villars F (1949) On the invariant regularization in relativistic quantum theory. Rev Mod Phys 21:434–444
47. Wilson KG (1974) Confinement of quarks. Phys Rev D 10:2445–2459
48. Deur A, Brodsky SJ, de Teramond GF (2016) The QCD running coupling. Nucl Phys 90:1
49. Taylor JC (1971) Ward identities and charge renormalization of the Yang-Mills field. Nucl Phys B 33:436–444
50. Slavnov AA (1972) Ward identities in gauge theories. Theor Math Phys 10:99–107
51. Ward JC (1950) An identity in quantum electrodynamics. Phys Rev 78:182–182
52. Takahashi Y (1957) On the generalized Ward identity. Nuovo Cim 6:371
53. Gross DJ (1975) Applications of the renormalization group to high-energy physics. Conf Proc C 7507281:141–250
54. Gross DJ, Wilczek F (1973) Ultraviolet behavior of nonabelian gauge theories. Phys Rev Lett 30:1343–1346 (Taylor JC (ed))
55. Gross DJ, Wilczek F (1973) asymptotically free gauge theories—I. Phys Rev D 8:3633–3652
56. Politzer HD (1973) Reliable perturbative results for strong interactions? Phys Rev Lett 30:1346–1349 (Taylor JC (ed))
57. Jones D (1974) Two-loop diagrams in Yang-Mills theory. Nucl Phys B 75:531–538. ISSN: 0550-3213

58. Caswell WE (1974) Asymptotic behavior of nonabelian gauge theories to two loop order. Phys Rev Lett 33:244
59. Egorian E, Tarasov OV (1979) Two loop renormalization of the QCD in an arbitrary gauge. Teor Mat Fiz 41:26–32
60. Egoryan ÉS, Tarasov OV (1979) Renormalization of quantum chromodynamics in the two-loop approximation in an arbitrary gauge. Theor Math Phys 41:863
61. Tarasov OV, Vladimirov AA, Zharkov AY (1980) The Gell-Mann-Low function of QCD in the three loop approximation. Phys Lett B 93:429–432
62. Larin SA, Vermaseren JAM (1993) The three loop QCD beta function and anomalous dimensions. Phys Lett B 303:334–336
63. Chetyrkin KG, Kuhn JH, Steinhauser M (2000) RunDec: a mathematica package for running and decoupling of the strong coupling and quark masses. Comput Phys Commun 133:43–65
64. Van Ritbergen T, Vermaseren J, Larin S (1997) The four loop beta function in quantum chromo dynamics. Phys Lett B 400:379–384
65. Czakon M (2005) The four-loop QCD beta-function and anomalous dimensions. Nucl Phys B 710:485–498
66. Baikov PA, Chetyrkin KG, Kuhn JH (2017) Five-loop running of the QCD coupling constant. Phys Rev Lett 118:082002
67. Chetyrkin KG, Kniehl BA, Steinhauser M (1997) Strong coupling constant with flavor thresholds at four loops in the MS scheme. Phys Rev Lett 79:2184–2187
68. Alekhin S, Blumlein J, Moch S, Placakyte R (2017) Parton distribution functions, α_S and heavy quark masses for LHC Run II. Phys Rev D 96:014011
69. Khachatryan V et al (2017) Measurement and QCD analysis of double-differential inclusive jet cross sections in pp collisions at $\sqrt{s} = 8$ TeV and cross section ratios to 2.76 and 7 TeV. JHEP 03:156
70. CMS physics results combined. Accessed 17 Jan 2022. https://twiki.cern.ch/twiki/bin/view/CMSPublic/PhysicsResultsCombined
71. Behnke O, Geiser A, Lisovyi M (2015) Charm, beauty and top at HERA. Prog Part Nucl Phys 84:1–72
72. Abdallah J et al (2008) Study of b-quark mass effects in multijet topologies with the DELPHI detector at LEP. Eur Phys J C 55:525–538
73. Abdallah J et al (2006) Determination of the b quark mass at the M(Z) scale with the DELPHI detector at LEP. Eur Phys J C 46:569–583
74. Barate R et al (2000) A measurement of the b quark mass from hadronic Z decays. Eur Phys J C 18:1–13
75. Abbiendi G et al (2001) Determination of the b quark mass at the Z mass scale. Eur Phys J C 21:411–422
76. Brandenburg A, Burrows PN, Muller D, Oishi N, Uwer P (1999) Measurement of the running b quark mass using $e^+ e^- \to$ b anti-b g events. Phys Lett B 468:168–177
77. Abramowicz H et al (2014) Measurement of beauty and charm production in deep inelastic scattering at HERA and measurement of the beauty-quark mass. JHEP 09:127
78. Gizhko A et al (2017) Running of the charm-quark mass from HERA deep-inelastic scattering data. Phys Lett B 775:233–238
79. Sirunyan AM et al (2020) Running of the top quark mass from proton-proton collisions at $\sqrt{s} = 13$ TeV. Phys Lett B 803:135263
80. Bigi IIY, Shifman MA, Uraltsev NG, Vainshtein AI (1994) The Pole mass of the heavy quark: perturbation theory and beyond. Phys Rev D 50:2234–2246
81. Beneke M, Braun VM (1994) Heavy quark effective theory beyond perturbation theory: renormalons, the pole mass and the residual mass term. Nucl Phys B 426:301–343
82. Smith MC, Willenbrock SS (1997) Top quark pole mass. Phys Rev Lett 79:3825–3828
83. Tarrach R (1981) The pole mass in perturbative QCD. Nucl Phys B 183:384–396
84. Tarasov OV (2020) Anomalous dimensions of quark masses in the three-loop approximation. Phys Part Nucl Lett 17:109–115

85. Larin SA (1993) The renormalization of the axial anomaly in dimensional regularization. Phys Lett B 303:113–118
86. Chetyrkin KG (1997) Quark mass anomalous dimension to $\mathcal{O}(\alpha_S^4)$. Phys Lett B 404:161–165
87. Vermaseren JAM, Larin SA, van Ritbergen T (1997) The four loop quark mass anomalous dimension and the invariant quark mass. Phys Lett B 405:327–333
88. Hoang AH, Stewart IW (2008) Top mass measurements from jets and the Tevatron top-quark mass: mass anomalous dimension. Nucl Phys B Proc Suppl 185:220–226 (Bellettini G, Chiarelli G, Tenchini R (eds))
89. Hoang AH (2014) The top mass: interpretation and theoretical uncertainties in 7th international workshop on top quark physics. arXiv: 1412.3649 [hep-ph]
90. Hoang AH et al (2018) The MSR mass and the $\mathcal{O}(\Lambda_{\text{QCD}})$ renormalon sum rule. JHEP 04:003
91. Han MY, Nambu Y (1965) Three-triplet model with double SU(3) symmetry. Phys Rev 139:B1006–B1010
92. Aitchison I, Hey T (2003) Gauge theories in particle physics: a practical introduction, vol 1, From relativistic quantum mechanics to QED. http://www-spires.fnal.gov/spires/find/books/www?cl=QC793.3.F5A34::2012
93. Ellis R, Ross D, Terrano A (1981) The perturbative calculation of jet structure in $e^+ e^-$ annihilation. Nucl Phys B 178:421–456. ISSN: 0550-3213
94. Brandelik R et al (1979) Evidence for planar events in e +e? annihilation at high energies. Phys Lett B 86:243–249. ISSN: 0370-2693
95. Abbiendi G et al (2001) A simultaneous measurement of the QCD color factors and the strong coupling. Eur Phys J C 20:601–615
96. Catani S, Dokshitzer YL, Seymour MH, Webber BR (1993) Longitudinally invariant K_t clustering algorithms for hadron hadron collisions. Nucl Phys B 406:187–224
97. Dokshitzer Y, Leder G, Moretti S, Webber B (1997) Better jet clustering algorithms. J High Energy Phys 1997:001–001
98. Cacciari M, Salam GP, Soyez G (2008) The anti-kT jet clustering algorithm. JHEP 04:063
99. Skands P, Carrazza S, Rojo J (2014) Tuning PYTHIA 8.1: the Monash 2013 tune. Eur Phys J C 74:3024
100. Gustafson G, Pettersson U, Zerwas P (1988) Jet final states in WW pair production and colour screening in the QCD vacuum. Phys Lett B 209:90–94. ISSN: 0370-2693
101. Argyropoulos S, Sjostrand T (2014) Effects of color reconnection on $t\bar{t}$ final states at the LHC. JHEP 11:043
102. Christiansen JR, Skands PZ (2015) String formation beyond leading colour. JHEP 08:003
103. Dittmaier S, Huss A, Speckner C (2012) Weak radiative corrections to dijet production at hadron colliders. JHEP 11:095
104. Abe F et al (1995) Observation of top quark production in $p\bar{p}$ collisions. Phys Rev Lett 74:2626–2631
105. Abachi S et al (1995) Observation of the top quark. Phys Rev Lett 74:2632–2637
106. Sirunyan AM et al (2019) Measurement of the top quark polarization and $t\bar{t}$ spin correlations using dilepton final states in proton-proton collisions at $\sqrt{s} = 13$ TeV. Phys Rev D 100:072002
107. Aad G et al (2013) Measurement of the top quark charge in pp collisions at $\sqrt{s} = 7$ TeV with the ATLAS detector. JHEP 11:031
108. Tumasyan A et al (2022) Measurement and QCD analysis of double-differential inclusive jet cross sections in pp collisions at $\sqrt{s} = 13$ TeV. JHEP 02:142
109. De Jong P (2009) Top physics at the LHC. arXiv: 0902.4798 [hep-ex]
110. Collaboration D (2011) Useful diagrams of top signals and backgrounds. Accessed 19 July 2022. https://www-d0.fnal.gov/Run2Physics/top/top_public_web_pages/top_feynman_diagrams.html
111. Campbell J, Neumann T, Sullivan Z (2021) Single-top-quark production in the t-channel at NNLO. JHEP 02:040
112. Kant P et al (2015) HatHor for single top-quark production: Updated predictions and uncertainty estimates for single top-quark production in hadronic collisions. Comput Phys Commun 191:74–89

113. Kieseler J, Lipka K, Moch S-O (2016) Calibration of the top-quark Monte Carlo mass. Phys Rev Lett 116:162001
114. Hoang AH, Plätzer S, Samitz D (2018) On the cutoff dependence of the quark mass parameter in angular ordered Parton showers. JHEP 10:200
115. Buckley A et al (2011) General-purpose event generators for LHC physics. Phys Rept 504:145–233
116. Butenschon M et al (2016) Top quark mass calibration for Monte Carlo event generators. Phys Rev Lett 117:232001
117. A precise interpretation for the top quark mass parameter in ATLAS Monte Carlo simulation (2021)
118. Abazov VM et al (2016) Measurement of the Inclusive $t\bar{t}$ production cross section in $p\bar{p}$ collisions at $\sqrt{s} = 1.96$ TeV and determination of the top quark pole mass. Phys Rev D 94:092004
119. Abazov VM et al (2011) Determination of the pole and $\bar{M}S$ masses of the top quark from the $t\bar{t}$ cross section. Phys Lett B 703:422–427
120. Aad G et al (2014) Measurement of the $t\bar{t}$ production cross-section using $e\mu$ events with b-tagged jets in pp collisions at $\sqrt{s} = 7$ and 8 TeV with the ATLAS detector. Eur Phys J C 74:3109 [Addendum: Eur Phys J C 76:642 (2016)]
121. Khachatryan V et al (2016) Measurement of the $t\bar{t}$ production cross section in the $e - \mu$ channel in proton-proton collisions at $\sqrt{s}$ =7 and 8 TeV. JHEP 08:029
122. Chatrchyan S et al (2014) Determination of the top-quark pole mass and strong coupling constant from the $t\bar{t}$ production cross section in pp collisions at $\sqrt{s} = 7$ TeV. Phys Lett B 728:496–517 [Erratum: Phys Lett B 738:526–528 (2014)]
123. LHC Top Working Group (2022) LHCTopWG summary plots. Accessed 14 July 2022. https://twiki.cern.ch/twiki/bin/view/LHCPhysics/LHCTopWGSummaryPlots
124. Garzelli MV, Kemmler L, Moch S, Zenaiev O (2021) Heavy-flavor hadro-production with heavy quark masses renormalized in the $\bar{M}S$, MSR and on-shell schemes. JHEP 04:043
125. Sirunyan AM et al (2020) Measurement of $t\bar{t}$ normalised multi- differential cross sections in pp collisions at $\sqrt{s} = 13$ TeV, and simultaneous determination of the strong coupling strength, top quark pole mass, and Parton distribution functions. Eur Phys J C 80:658
126. H1 and ZEUS Collaborations (2015) Combination of measurements of inclusive deep inelastic e±p scattering cross sections and QCD analysis of HERA data. Eur Phys J C 75:580
127. Anderson M et al (2019) Search for invisible modes of nucleon decay in water with the SNO+ detector. Phys Rev D 99:032008
128. Abe K et al (2017) Search for proton decay via $p \to e^+\pi^0$ and $p \to \mu^+\pi^0$ in 0.31 megaton: years exposure of the Super-Kamiokande water Cherenkov detector. Phys Rev D 95:012004
129. Gell-Mann M (1964) A schematic model of baryons and mesons. Phys Lett 8:214–215. ISSN: 0031-9163
130. Zweig G (1964) An SU(3) model for strong interaction symmetry and its breaking. Version 1
131. Zweig G (1964) In: Lichtenberg DB, Rosen SP (eds) Developments in the quark theory of hadrons, vol 1, 1964–1978, pp 22–101
132. Bjorken JD, Paschos EA (1969) Inelastic electron proton and gamma proton scattering, and the structure of the nucleon. Phys Rev 185:1975–1982
133. Taylor JC (1974) Photon-hadron interactions. Phys Bull 25:105–105
134. Miller G et al (1972) Inelastic electron-proton scattering at large momentum transfers and the in elastic structure functions of the proton. Phys Rev D 5:528–544
135. Eichten T et al (1973) Measurement of the neutrino-nucleon and antineutrino-nucleon total cross sections. Phys Lett B 46:274–280. ISSN: 0370-2693
136. Collins JC, Soper DE, Sterman GF (1989) Factorization of hard processes in QCD. Adv Ser Direct High Energy Phys 5:1–91
137. Data analysis in high energy physics: a practical guide to statistical methods (2013). In: Behnke O, Kröninger K, Schörner-Sadenius T, Schott G (eds). Wiley-VCH, Weinheim, Germany. ISBN: 978-3-527-41058-3, 978-3-527-65344-7, 978-3-527-65343-0

138. Gribov VN, Lipatov LN (1972) Deep inelastic ep scattering in perturbation theory. Sov J Nucl Phys 15:438
139. Altarelli G, Parisi G (1977) Asymptotic freedom in Parton language. Nucl Phys B 126:298. ISSN: 0550-3213
140. Curci G, Furmanski W, Petronzio R (1980) Evolution of Parton densities beyond leading order: the non-singlet case. Nucl Phys B 175:27. ISSN: 0550-3213
141. Furmanski W, Petronzio R (1980) Singlet Parton densities beyond leading order. Phys Lett B 97:437. ISSN: 0370-2693
142. Moch S, Vermaseren JAM, Vogt A (2004) The three-loop splitting functions in QCD: the nonsinglet case. Nucl Phys B 688:101
143. Vogt A, Moch S, Vermaseren JAM (2004) The three-loop splitting functions in QCD: the singlet case. Nucl Phys B 691:129
144. Alekhin SI et al (2005) DGLAP evolution and Parton fits
145. Ellis RK, Stirling WJ, Webber BR (2003) QCD and collider physics Photography by S. Vascotto. Cambridge University Press, Cambridge
146. Peskin ME, Schroeder DV An introduction to quantum field theory. ISBN: 9780201503975, 0201503972. http://www.slac.stanford.edu/spires/find/books/www?cl=QC174.45
147. Donnachie S, Dosch HG, Nachtmann O, Landshoff P (2004) Pomeron physics and QCD 978–0, ISBN: 978-0-511-06050-2-521-78039-1, 978-0-521-67570-3. Cambridge University Press, Cambridge
148. Abramowicz H et al (2015) Combination of measurements of inclusive deep inelastic $e \pm p$ scattering cross sections and QCD analysis of HERA data. Eur Phys J C 75:580
149. Abe F et al (1996) Inclusive jet cross section in $\bar{p}p$ collisions at $\sqrt{s} = 1.8$ TeV. Phys Rev Lett 77:438–443
150. Abulencia A et al (2007) Measurement of the inclusive jet cross section using the kT algorithm in $p\bar{p}$ collisions at $\sqrt{s} = 1.96$ TeV with the CDF II detector. Phys Rev D 75:092006 [Erratum: Phys Rev D 75:119901 (2007)]
151. Stirling J (2022) 13 TeV LHC x, Q^2 parton kinematics Private communication. Accessed 07 Apr 2022. http://www.hep.ph.ic.ac.uk/~wstirlin/plots/plots.html

Chapter 3
Effective Field Theory

This chapter is dedicated to a discussion on the effective field theory (EFT) approach. Effective theories provide a simplified description of the investigated system, and have applications in investigating both SM and BSM physics. An EFT is designed to be robust within a given region of validity, and it may contain additional symmetries not present in the full theory [1]. They are particularly useful for multiscale problems, where the full theory can be approximated by an effective theory at a scale where all the features of the full theory are not necessary for providing predictions of an observable. In an EFT, a complicated or unknown mechanism such as the exchange of a new particle, is replaced by a computation requiring no detailed assumptions of the mechanism included in the full theory. The EFT produces approximate results that can eventually be matched to the full theory in the regime for which the EFT is designed. A historical example of this is given by Fermi's 1934 theory for weak decays [2], which is an EFT valid at low energies. This EFT was eventually UV-completed by the SM, which described the observed decays in terms of interactions mediated by the W boson. The power of Fermi's EFT was then in permitting computations of weak decays using the experimentally determined Fermi constant, without a description of the dynamics of the weak bosons.

Fermi's theory is an instance of the type of EFTs that correspond to removing heavy degrees of freedom from a full theory, as the W bosons are not explicitly included in its description of the weak decay, but shrunk into a point in the Feynman diagram:

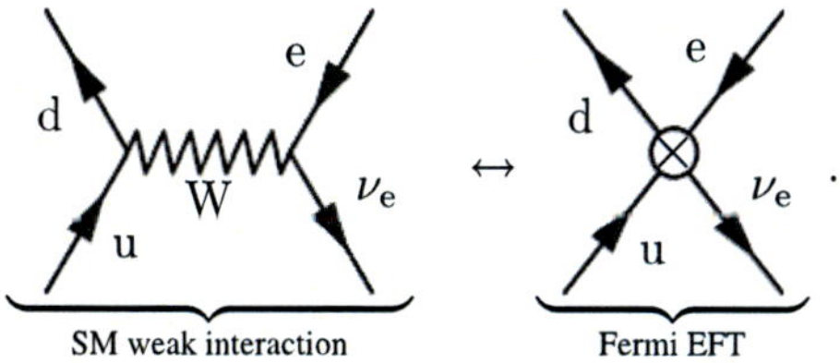

(3.1)

T. Mäkelä, *Towards Global Interpretation of LHC Data*, Springer Theses,
https://doi.org/10.1007/978-3-031-29779-3_3

The Standard Model effective field theory (SMEFT) is also an EFT of this kind. It offers a model-independent description of effective corrections to processes involving only SM particles in the initial and final states. Another type of EFTs are those constructed by removing only parts of the fields. One instance of such an EFT is the soft-collinear effective theory (SCET) [3–10], which has led to novel factorization theorems for cross section predictions. These two types of EFTs are used in this thesis, but also other types exist; e.g. the heavy-quark effective theory [11, 12] is based on restricting the dynamics of heavy particles.

The SMEFT is discussed in Sect. 3.1. Section 3.2 focuses on the 4-quark contact interactions (CI) appearing in SMEFT at dimension-6, and details the theory behind the CIJET calculation [13, 14] utilized in this thesis. Furthermore, the NLL-JET calculation used in the thesis is based on a factorization theorem in the context of SCET, which is discussed in Sect. 3.3.

3.1 The Standard Model Effective Field Theory

A theorem by Steven Weinberg states that for a given set of asymptotic states, the most general S-matrix is obtained from a Lagrangian constructed out of all terms allowed by the symmetries of the theory. This results in the most general matrix elements to any order in perturbation theory. It is then consistent with the assumed symmetries, analyticity, cluster decomposition, and perturbative unitarity [15]. For the sake of renormalizability, the operators in the SM Lagrangian are restricted to dimension 4. However, this requirement is dropped in SMEFT, and the idea of Weinberg's theorem is applied by extending the SM Lagrangian with higher-dimensional operators composed of the SM fields, so that the Lagrangian has the form

$$\mathscr{L}_{\text{SMEFT}} = \mathscr{L}_{\text{SM}} + \sum_{d \geq 5} \frac{1}{\Lambda_d^{d-4}} \sum_n c_n^{(d)} O_n^{(d)} \tag{3.2}$$

where Λ is the energy scale of the new interactions, the sum over n runs over the combinations of SM fields that result in the dimension-d composite operators O respecting the symmetries of the SM, and the *Wilson coefficient* c_n is analogous to a coupling [1].

If the scale of new physics is close to the EW scale of $v \approx 246\,\text{GeV}$, it can be probed by direct searches at LHC energies. In this case, the standard SMEFT approach could be suboptimal since an expansion parameter of $\sim v/\Lambda$ may lead to large contributions. The formalism of SMEFT can however be used also in this case, if also exotic fields are included in the non-renormalizable operators. This can parametrize for instance the interactions between the SM fields and weakly-interacting massive particles representing dark matter, even if the exotic particles' masses would be close to the EW scale [16–18]. Moreover, there are also low-energy effective field theories for physics below the EW scale, formed by integrating out heavy SM degrees of freedom [19].

Furthermore, if Λ is beyond the energy scale of the experiment, the effects of new physics can be visible in an indirect search looking for deviations in the continuous spectra from the SM predictions. However, an indirect signal has an ambiguity arising from the model dependencies. When matching onto an effective theory at lower energies, the non-analytic structure of correlation functions due to heavy states is projected out as formalized by the decoupling theorem [20]. The description of such physics is then a well-suited task for SMEFT: in the regime where the BSM exchanges only appear in virtual loops, they can be integrated out in the path integral. As an illustration, if a full theory has the heavy field ψ_h and the light fields ψ_l, the effective Lagrangian $\mathscr{L}_{\text{EFT}}$ is obtained from the effective action S_{EFT} via [21, 22]

$$\begin{aligned} e^{iS_{\text{EFT}}} &= \int (D\psi_h)(D\psi_h^\dagger) \exp\left(i \int d^4x \mathscr{L}(\psi_l, \psi_h)\right) \\ &= \exp\left(i \int d^4x \left(\mathscr{L}(\psi_l, 0) + O(\psi_l)\right)\right) = \exp\left(i \int d^4x \mathscr{L}_{\text{EFT}}(\psi_l)\right) \end{aligned} \tag{3.3}$$

where $O(\psi_l)$ is an effective composite operator constructed out of the light fields. The mass dimension of O may well be higher than 4, making it nonrenormalizable.

As a consequence of allowing for nonrenormalizable operators, a Feynman diagram's superficial degree of divergence in 4-dimensional spacetime becomes [23]

$$D = 4 - \frac{3}{2}n_F - n_B - \sum_{d \geq 5} n_V^{(d)} [\Lambda^{4-d}] \tag{3.4}$$

where n_F (n_B) is the number of fermions (bosons) and $n_V^{(d)}$ the number of vertices corresponding to the nonrenormalizable operator with a coupling of dimension $[\Lambda^{d-4}]$. Therefore, including operators with $[\Lambda^{4-d}] < 0$ and not constraining $n_V^{(d)}$ would lead to infinite numbers of possible values for n_F and n_B, and consequently infinitely many Green's functions for which some 1PI graphs have positive degrees of divergence. Therefore, the full renormalization of such a theory would require an infinite number of counterterms [23]. However, nonrenormalizable theories can be renormalized *at a given order* by adding all the possible terms consistent with the symmetries of the theory [23]. Counterterms are then determined by computing sufficiently many physical quantities, and fixing them to external data, or by matching the computations to some other theory in a given limit. Once the effective coefficients are known at a high scale, their values at lower scales are determined via an expansion in terms of the relevant SM coupling constant, or by RGE evolution [1].

At dimension 5, the only operator satisfying the symmetry requirements of the SM is the *Weinberg operator* [18, 24]

$$O^{(5)} = \epsilon_{ij}\epsilon_{km} H^i H^k L^j C L^m \tag{3.5}$$

where ϵ is the antisymmetric tensor, C is the charge conjugation operator and H and L are the Higgs and lepton doublets, respectively. Such an operator can be used for introducing (Majorana) neutrino mass terms, and results in lepton number violation at a scale Λ_5. However, this does not need to equal the scale of higher-dimensional operators in the SMEFT Lagrangian. If the Wilson coefficient was assumed one, the generation of neutrino masses below 1 eV would imply $\Lambda_5 > 10^{14}$ GeV [1]. There is no lepton number violation in the SM, and it is not required to be a part of SMEFT, either: if $\Lambda_6 \ll \Lambda_5$, dimension six operators dominate over dimension five [19]. Many BSM effects are therefore parametrized using dimension six operators [1].

The first attempts to collect a complete set of all possible $SU(3) \times SU(2) \times U(1)$ invariant dimension-6 operators were carried out in Refs. [25, 26]. However, there are several alternative ways to define the complete set of non-redundant operators, giving rise to the choice of an *operator basis* [27, 28]. It was noticed in Ref. [27], that some of the operators in Ref. [26] vanish via EOMs, and the authors present a list of the operators whose linear combinations do not vanish due to EOMs, thereby introducing the widely-used Warsaw basis. A full renormalization of the dimension-6 Warsaw basis has been performed at 1-loop using DR and the $\overline{\text{MS}}$ scheme, leading to the determination of the full 2499×2499 anomalous dimension matrix [29–34]. Moreover, dimension 7 operators have been collected e.g. in [35], and dimension 8 operators in [36]. However, the question of which operators to choose for a given problem depends on the SM process to which the effective corrections are to be computed.

3.2 Models of Quark Substructure and 4-Quark Contact Interactions

In the SM, the quarks and leptons are assumed fundamental point-like particles. However, there is no evident explanation why three fermion generations exist, with very similar properties except for the particle masses. Attempts to economize the amount of fundamental particles have lead to theories where the SM fermions are not elementary but composite particles with hypothetical constituents referred to as *preons*. Theories describing quarks as composite particles [37–39] postulate inter-preon interactions at scales much higher than the masses of the SM quarks, whereas at energies well below Λ the interactions are approximated by *contact interactions* (CI) connecting four fermion legs to the same vertex. As done in this thesis, the CI can be described via dimension-6 SMEFT operators.

Analogous to QCD, where quarks are confined within a hadron's radius proportional to $1/\Lambda_{\text{QCD}}$, the preons are postulated to carry a *hypercolor* (HC) leading to non-Abelian interactions and confining them to within the radii of the leptons and quarks. In this case, the scale of hypercharge Λ_{HC} could be much greater than the QCD scale Λ_{QCD}, even if the masses of the individual preons were small [40].

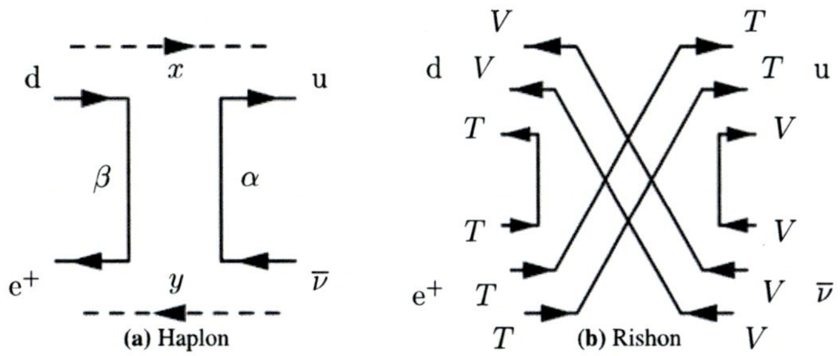

Fig. 3.1 Example diagrams for the $\mathrm{de}^+ \to \mathrm{u}\overline{\nu}$ process in the **a** Haplon and **b** Rishon models. The solid (dashed) lines represent fermionic (scalar) preons [40]

If the quarks and leptons are composite fermions, they could be constituted by various combinations of bosonic and leptonic preons [40]. The simplest model, the *Haplon model* [41], postulates a colorless fermion doublet (α, β), and a quartet of scalars (x^i, y) where i runs over color. The first generation of SM fermions is then constructed as $\mathrm{d} = \beta x, \mathrm{u} = \alpha x, \mathrm{e}^- = \alpha y$ and $\nu = \alpha y$. Several variations of the Haplon model can be achieved by describing more than one of the preon types as color triplets. The Haplon models also contain hypercolor- and color singlet bosons, which could possibly be identified as Higgs scalars or the weak vector bosons [40, 42]. The fundamental bosons of the theory are then taken to be the massless bosons and the possible *hypergluons* [40]. The higher-mass fermion generations are then explained as excitations of the first generation or as composites containing the same valence preons as the first generation with additional hypercolorless preon-antipreon pairs [40].

Alternatively, the *Rishon models* [43, 44] postulate only fermionic preons called rishons. The simplest such model involves only two rishons labeled T and V, carrying electrical charges of $e/3$ and 0, respectively. The first generation SM fermions would then be constituted as $\overline{\mathrm{d}} = TVV, \mathrm{u} = TTV, \mathrm{e}^+ = TTT$ and $\nu = VVV$. Further variations of the Rishon model make different assumptions on the amount of preons. For instance, Ref. [37] has three types of preons, and an analysis of all possible symmetry groups has been made in Ref. [45]. However, the possible solutions involving high rank groups would require large numbers of preons, hence not necessarily economizing the amount of elementary particles [40]. The physics of the Haplon and Rishon models are illustrated in Fig. 3.1, showing possible diagrams for the weak process $\mathrm{de}^+ \to \mathrm{u}\overline{\nu}$ in the context of both models.

However, the preon models are not free of problems. A model should address the mixing of quark generations without introducing lepton mixing. The preons can also serve as the ingredients for right handed neutrinos, and curing this may require e.g. introducing separate left-handed doublets and right-handed singlets, so that the weak bosons only couple to the left-handed particles, as in the Pati-Salam model [46]. Furthermore, certain processes forbidden in the SM would be permitted if the quarks and leptons consist of common preons [40]. As a particular example, the

Rishon model allows the decay of the proton via the $uu \rightarrow e^+\overline{d}$ process. Nonetheless, it has been argued that this decay would be chirally suppressed due to requiring a change of helicity [47]. In general, the models where leptons and quarks have common constituents tend to require either a large scale Λ_{HC} or special mechanisms for suppressing unobserved decays [40].

It was estimated in Ref. [48] that the Tevatron experiments can probe new physics scales from 2 to 5 TeV, and the LHC experiments in the range of 15–20 TeV. If the scale is at such a level, the CI are required to be flavor symmetric at least for the first two generations of quarks, to avoid large flavor-changing neutral current interactions [48]. If the preons are color-charged, quark scattering processes could be mediated either by gluon exchange, or by a rearrangement of the substructure particles. The interpreon interaction strength is expected to become of the order of unity as the preons approach one another within distances below $1/\Lambda_{\text{HC}}$ [40]. This leads to a modification of the cross section that can be tested using jet production as done in this thesis, where the considered CI are taken to be flavor symmetric products of electroweak isoscalar quark currents [13].

For the production of jets at a hadron collider, the effective corrections to quark scattering are introduced in SMEFT via dimension-6 operators. The Lagrangian reads [13, 14]

$$\mathscr{L}_{\text{SM+CI}} = \mathscr{L}_{\text{SM}} + \frac{2\pi}{\Lambda^2} \sum_n c_n O_n, \tag{3.6}$$

where the factor of 2π is due to the convention chosen in Ref. [49]. The dimension 6 operators O_n given in the chiral basis by

$$O_1 = \delta^{ij} \overline{q}_{Lci} \gamma_\mu q^c_{Lj} \delta^{kl} \overline{q}_{Ldk} \gamma^\mu q^d_{Ll}, \tag{3.7}$$

$$O_2 = (T_a)^{ij} \overline{q}_{Lci} \gamma_\mu q^c_{Lj} (T^a)^{kl} \overline{q}_{Ldk} \gamma^\mu q^d_{Ll}, \tag{3.8}$$

$$O_3 = \delta^{ij} \overline{q}_{Lci} \gamma_\mu q^c_{Lj} \delta^{kl} \overline{q}_{Rdk} \gamma^\mu q^d_{Rl}, \tag{3.9}$$

$$O_4 = (T_a)^{ij} \overline{q}_{Lci} \gamma_\mu q^c_{Lj} (T^a)^{kl} \overline{q}_{Rdk} \gamma^\mu q^d_{Rl}, \tag{3.10}$$

$$O_5 = \delta^{ij} \overline{q}_{Rci} \gamma_\mu q^c_{Rj} \delta^{kl} \overline{q}_{Rdk} \gamma^\mu q^d_{Rl}, \tag{3.11}$$

$$O_6 = (T_a)^{ij} \overline{q}_{Rci} \gamma_\mu q^c_{Rj} (T^a)^{kl} \overline{q}_{Rdk} \gamma^\mu q^d_{Rl}, \tag{3.12}$$

where the sums in c and d run over generations, whereas i, j, k, l are color indices. The L and R subscripts denote the handedness of the quarks. Such operators lead to the vertices with 4 external quark legs, illustrated by the graphs in Fig. 3.2.

The CI investigated in this thesis correspond to either purely left-handed, vector-like, or axial vector-like exchanges. For these models, if the value of the Wilson coefficient c_1 multiplying the operator O_1 in Eq. (3.6) is fixed at the high scale Λ, the coefficients c_3 and c_5 are determined from c_1 according to how the handedness of quark lines may change through the CI, while the coefficients c_2, c_4 and c_6 are set to zero. Thus the effective corrections correspond to integrating out a color-singlet BSM exchange between the quark lines. In the left-handed singlet model, there are

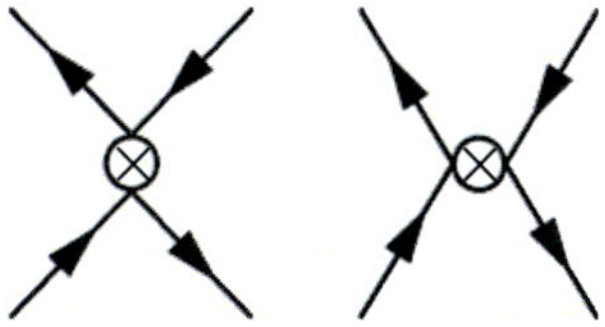

Fig. 3.2 The LO graphs for $q\bar{q} \to q\bar{q}$ scattering involving CI. The color indices are to be contracted between the quark legs connected to the same side of the vertex, making the two graphs distinct [14]

CI only between two left-handed lines, and hence $c_3 = c_5 = 0$. Vector-like and axial vector-like exchanges allow for interactions also between right-handed quarks, giving $c_5 = c_1$ in both cases. For interactions between quark lines of different handedness, the vector-like exchange implies $c_3 = 2c_1$, whereas the axial vector-like model has $c_3 = -2c_1$. Alongside models of quark compositeness, such color-singlet exchanges can dominate in models involving Z′ bosons [50].

The color-octet operators O_2, O_4 and O_6 do not contribute at leading order, as their coefficients are set to zero at the high scale Λ in all the considered models. However, at NLO they contribute to the renormalization of the CI due to the *operator mixing* caused by the QCD loop corrections to the CI vertex. The CI operators are renormalized using a 6×6 matrix Z in the relation $O_i^{\text{bare}} = Z^{ij} O_j$. Due to the mixing, all six operators must be taken into account in canceling the UV divergences, and the values of c_2, c_4 and c_6 can be non-zero at scales μ below Λ.

The virtual NLO corrections involving CI and QCD vertices are illustrated for $qq' \to qq'$ scattering in Fig. 3.3. The IR divergences are expected to cancel with the real emission corrections shown in Fig. 3.4, which are computed in the CIJET calculation using the phase space slicing based two cutoff method [51] in which the phase space regions containing soft and collinear singularities are separated from non-singular regions by two cut off parameters, and a cross-check is performed with the subtraction based dipole method [52]. In addition to the real and virtual vertex corrections, the NLO cross section computation includes the penguin graphs shown in Fig. 3.5 [14].

In the CIJET fixed-order calculation of double-differential inclusive jet production cross sections, in bins of rapidity y and transverse momentum p_T, the coefficients $c_i(\mu)$ are obtained from an expansion up to $\mathcal{O}(\alpha_S)$, and the cross section is truncated to $c_1(\mu = \Lambda)(1 + \mathcal{O}(\alpha_S))$. The color-singlet CI contribution to the differential cross section in each bin b then has the form

$$\sigma_b = \sum_{n\in\{1,3,5\}} \frac{c_n}{\Lambda^2}\left(b_n + a_n \log\left(\frac{\Lambda}{\mu_0}\right)\right) + \sum_{n\in\{1,3,5\}} \frac{c_n^2}{\Lambda^4}\left(b_{nn} + a_{nn} \log\left(\frac{\Lambda}{\mu_0}\right)\right), \tag{3.13}$$

consisting of interference terms between CI and QCD, and of terms with the CI contribution squared. In this thesis, the reference scale μ_0 is set to the center of each p_{T} bin. The coefficients a_n, a_{nn}, b_n and b_{nn} are computed by the CIJET code, with

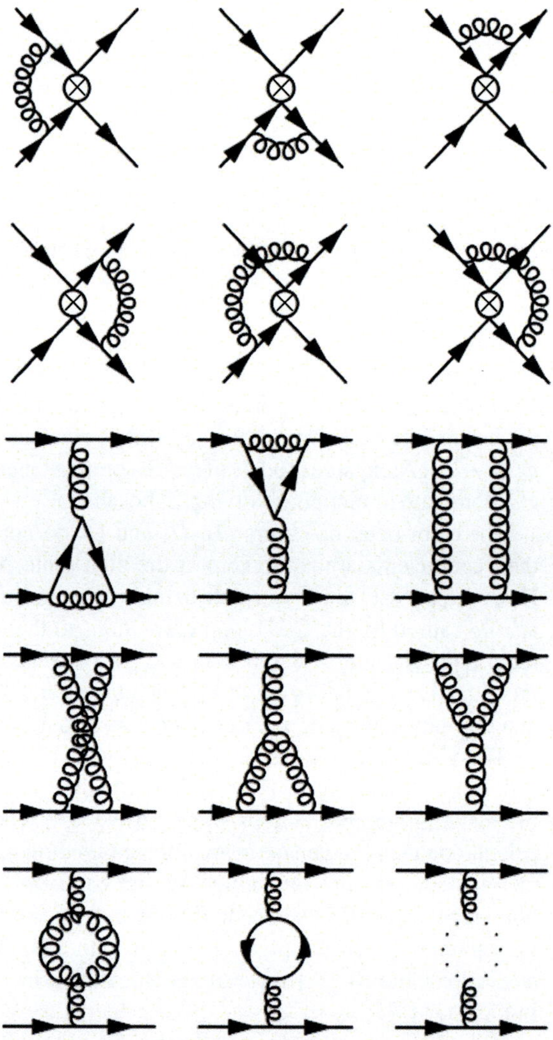

Fig. 3.3 The NLO virtual corrections to the CI vertex for the $qq' \to qq'$ process, as well as the SM QCD loop graphs [14]

the a_n and a_{nn} vanishing for all n at LO [13, 14]. The CI contribution is expected to appear as a deviation from the SM jet production cross section at high-p_{T} and low rapidity y. This is illustrated by the ratio of the SMEFT and SM cross sections in Fig. 3.6, for the left-handed and vector-like CI with $\Lambda = 50\,\mathrm{TeV}$ and $c_1 = \pm 1$ for constructive ($-$) and destructive ($+$) interference with the SM gluon exchange. With values of c_1/Λ^2 close to zero, the deviations from the SM would be small in all y and p_{T} regions, but Fig. 3.6 indicates that the effects of the considered CI are mostly present in the central rapidity region even at high values of c_1/Λ^2.

As an alternative to the fixed-order computation, the CI Wilson coefficients could be determined at $\mu < \Lambda$ by evolving them with an RGE accounting for the QCD

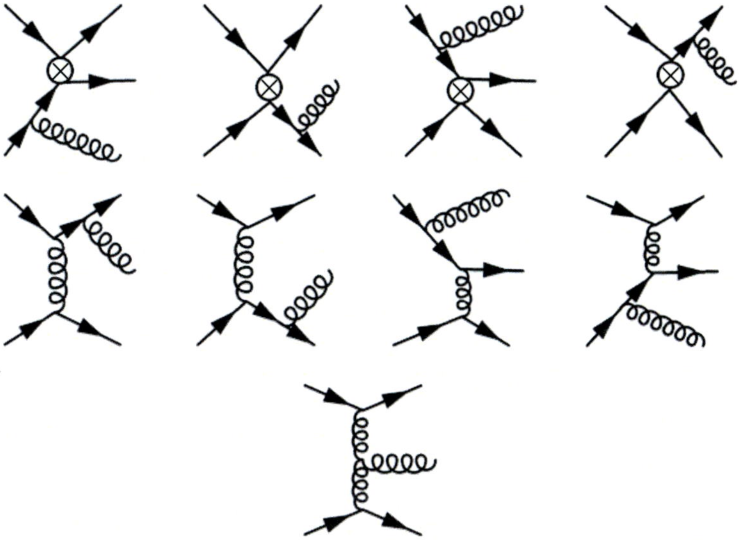

Fig. 3.4 NLO $qq' \to qq'$ real emission graphs involving CI, as well as the SM QCD graphs [14]

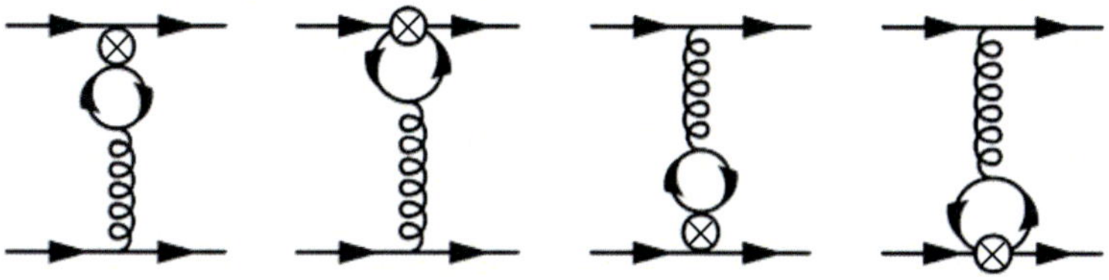

Fig. 3.5 The penguin diagrams involving CI that contribute to the $qq' \to qq'$ process at NLO [14]

running effects [53]. This has been considered for CI in dijet production in Refs. [14, 54]. However, the effect is small and is neglected in the single inclusive jet production computation. Furthermore, it should be noted that NLO results obtained using the EFT approach depend on the choice of scheme [1]. The CIJET program is based on NLO calculations carried out using the Feynman-'t Hooft gauge and the DR scheme in $d = 4 - 2\varepsilon$ dimensions with the naive γ_5 prescription, in which the γ_5 anticommutes with γ_μ, $\mu \in \{1, d\}$ [53, 55].

3.3 Soft-Collinear Effective Theory

Various processes at hadron colliders typically involve physics from high energy scales, such as the p_T of a jet, to a soft scale, such as the mass of the proton. Separating the effects at different scales to obtain factorization theorems is conventionally done at

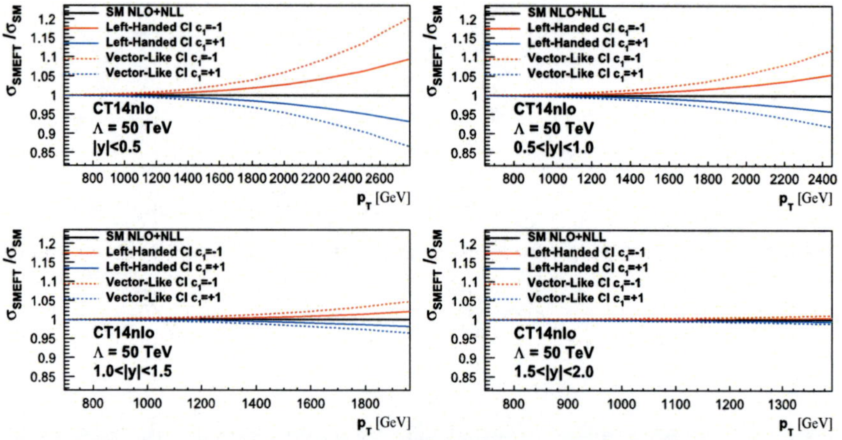

Fig. 3.6 The deviation from the SM jet cross section spectrum in pp collisions at $\sqrt{s} = 13\,\text{TeV}$, from purely left-handed and vector-like CI models, assuming the Wilson coefficient $c_1 = \pm 1$ and the new physics scale $\Lambda = 50\,\text{TeV}$, and using the CT14 PDF at NLO

the level of diagrams. However, working instead at the level of an effective Lagrangian may reveal consequences of gauge invariance that would not otherwise be manifest. The approach also provides possibilities for resumming large logarithms appearing in jet cross section computations by virtue of renormalization group evolution [56]. An EFT that has proven valuable in this regard is the soft-collinear effective theory (SCET), in which large energy-momentum transfers are either allowed or forbidden based on fixed spatial directions, such as those of jets. The theory can be motivated by considering the factorization of soft and collinear radiation at tree-level in QCD, so that at leading power e.g.

$$\langle X_1; \ldots ; X_m; X_S | \overline{q}_1 \ldots q_m \rangle \simeq \langle X_1 | \overline{q}_1 W_1 \rangle \ldots \langle X_m | W_m^\dagger q_m \rangle \langle X_S | Y_1 \ldots Y_m^\dagger \rangle \quad (3.14)$$

for a process where X_i, $i \in [1, m]$ contain gluons collinear to the ith jet and X_S the soft gluons [23]. The W and Y are *Wilson lines* for collinear and soft radiation, respectively. The Wilson lines are objects for comparing the field at two different locations, e.g. the origin 0 and an arbitrary point x, and in general have the form [23]

$$W(x) = P \exp\left(i g_s T_a v^\mu \int_0^\infty ds G_\mu^a(x^\nu + s v^\nu) e^{-\varepsilon s} \right),$$

where P is for path ordering and v^μ is a direction vector chosen e.g. anticollinearly to the object the Wilson line is associated with. It is then desirable to construct an EFT where

$$\langle X_1; \ldots ; X_m; X_S|\overline{q}_1 \ldots q_m\rangle_{\text{QCD}} \simeq$$
$$C\langle X_1; \ldots ; X_m; X_S|\overline{q}_1 W_1 Y_1^\dagger \ldots Y_m W_m^\dagger q_m\rangle_{\text{EFT}} \quad (3.15)$$

where the Wilson coefficient C does not depend on the soft or collinear scales, but only on the hard scale. Since the tree-level matrix elements agree on both sides of Eq. (3.15), the IR divergences in the loops on both sides are expected to agree [23].

Performing the derivations of factorization theorems at the level of fields brings universality, as the same objects then appear in several different factorization theorems [23]. In SCET, a superselection rule is imposed on the Lagrangian by labeling high-momentum particles according to the direction they are collinear to, i.e. the directions of jets. It is then assumed that there are no vertices implying interactions between particles collinear to different directions. Considering two directions c and $\overline{c}$ to which the fields can be collinear, the quark and gluon operators of QCD are divided into soft and collinear parts as

$$\begin{cases} G^\mu & \to G_c^\mu + G_{\overline{c}}^\mu + G_s^\mu \\ q & \to q_c + q_{\overline{c}} + q_s \end{cases} . \quad (3.16)$$

Here the subscripts c and $\overline{c}$ indicate which direction the field is collinear to, whereas s is for soft fields. The SCET Lagrangian reads

$$\mathscr{L}_{\text{SCET}} = \mathscr{L}_c + \mathscr{L}_{\overline{c}} + \mathscr{L}_s + \mathscr{L}_{c+s}, \quad (3.17)$$

where $\mathscr{L}_c$, $\mathscr{L}_{\overline{c}}$, and $\mathscr{L}_s$ are obtained by summing contributions of the form of the massless QCD Lagrangian

$$\mathscr{L}_{\text{QCD}} = -\frac{1}{4} G_{\mu\nu}^a G_a^{\mu\nu} + i\overline{q}\gamma^\mu D_\mu q \quad (3.18)$$

but replacing the quark and gluon operators with the respective soft or collinear field. Interactions between soft and collinear fields are introduced in $\mathscr{L}_{c+s}$, but only for combinations of soft particles and those collinear to one direction, in a way that preserves momentum when the momentum scales of soft and collinear particles are assumed to be of different orders of magnitude. The forbidden interaction vertices are illustrated in Fig. 3.7.

Fig. 3.7 Forbidden vertices in SCET. The dotted lines represent soft radiation, whereas the solid line represents a high-energetic particle in one direction and the dashed line a particle collinear to another direction. Note that the solid and dashed lines are interchangeable here [56]

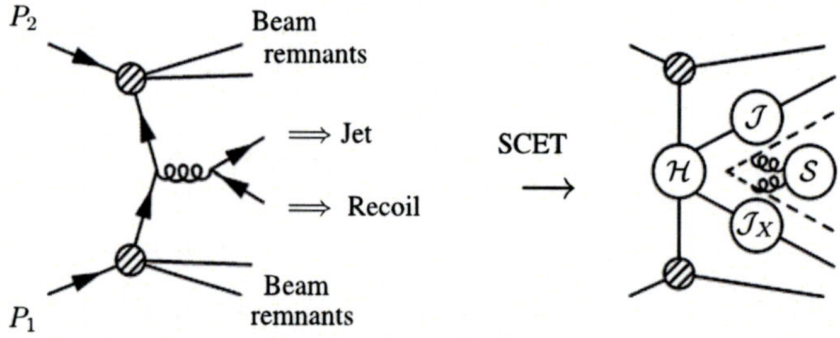

Fig. 3.8 In the context of SCET, the hard process in QCD single inclusive jet production is factorized into a hard function $\mathcal{H}$, soft function(s) $\mathcal{S}$, $\mathcal{J}$ for the signal jet and $\mathcal{J}_X$ for the recoil against the signal jet

Figure 3.8 illustrates a case where the cross section factorizes into a hard function $\mathcal{H}$, a signal jet function $\mathcal{J}$, a recoil function $\mathcal{J}_X$ and a soft function $\mathcal{S}$. The hard scattering event corresponds to $\mathcal{H}$, the outgoing partons to $\mathcal{J}$ and $\mathcal{J}_X$, while $\mathcal{S}$ accounts for soft radiation via soft Wilson lines. With a small parameter λ and a vector n^μ ($\overline{n}^\mu$) collinear (anticollinear) to the jet, and $\perp$ denoting the plane transverse to n^μ, the scaling of the integration momenta $k^\mu \doteq (n^\mu k_\mu, \overline{n}^\mu k_\mu, k_\perp)$ is proportional to $(1, 1, 1)$ in the hard region, $(\lambda^2, 1, \lambda)$ in the collinear region, $(1, \lambda^2, \lambda)$ in the anticollinear region and $(\lambda^2, \lambda^2, \lambda^2)$ in the soft region [56]. The hard, jet, recoil and soft functions contain logarithmic dependencies on their different characteristic scales, and the factorization allows for the functions to be evolved from these scales to a common renormalization scale μ_r, resumming the large logarithms in the process.

In SCET, the types of NLO diagrams with non-vanishing contributions to the hard function are

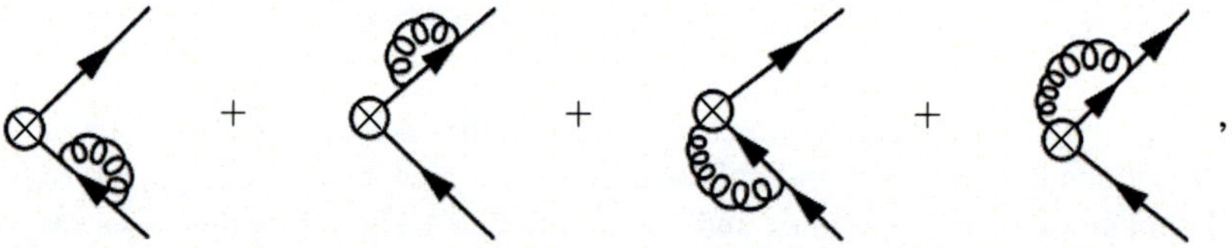

where the first two diagrams correspond to the QCD propagator corrections, and in the latter two the gluons connected to the operator vertex arise from an expansion of the Wilson line [23]. Regarding the jet functions, for instance the quark jet function can be obtained at NLO by evaluating Feynman diagrams

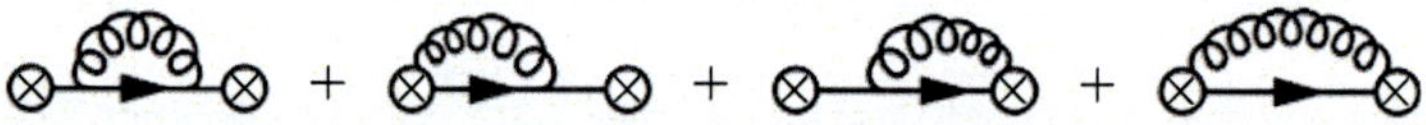

and proceeding by the cutting rules [23]. Alternatively, the jet function can be obtained as an integration of the splitting functions, which essentially describes the probability of a parton emitting collinear radiation [23]. The general soft function is based on the vacuum expectation value of the product of ordered soft Wilson lines. At NLO, any two of the four Wilson lines in this squared amplitude form may be connected by a soft gluon coming from expanding the exponential in the Wilson line, so that the non-vanishing contributions are given by

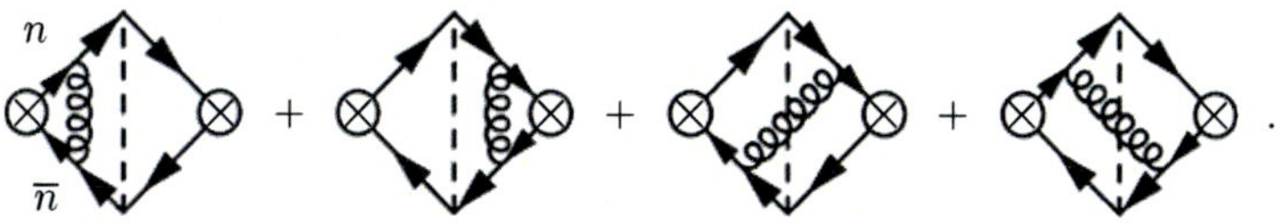

However, this is a simple picture, and precision computations such as Ref. [57] require separate collinear-soft functions and a global soft function. The *global soft function* accounts for wide-angle soft radiation and cannot resolve small jet radii, whereas the *collinear-soft function* accounts for radiation near the jet boundary and mixes with the jet function [57]. The two soft functions are realized by different operator definitions, as discussed in the following.

3.3.1 Joint Threshold and Jet-Radius Resummation for Inclusive Jet Production Cross Section

In perturbative computations at the order n, the threshold logarithms appear as $\alpha_S^n(\log^k(z)/z)$ for $k \leq 2n-1$, when $z = s_4/s$ with s (s_4) being the center of mass energy (invariant mass of partonic system recoiling against the observed jet). The threshold limit is then given by $z \to 0$. The logarithms on the jet radius $\mathcal{R}$ appear as $\alpha_S^n \log^k(\mathcal{R})$ for $k \leq n$ [58]. Reference [57] consistently addresses the logarithms by resumming them at next-to-leading logarithmic (NLL) accuracy to all orders in α_S.

The computation of Ref. [57], invoked in this thesis for improving the NLO inclusive jet production cross section to NLO+NLL, relies on a factorization formula derived within SCET. The total cross section reads

$$\frac{p_{\mathrm{T}}^2 d^2\sigma}{dp_{\mathrm{T}}^2 dy} = \sum_{i,j} \int_0^{V(1-W)} dz \int_{\frac{VW}{1-z}}^{1-\frac{1-V}{1-z}} dv\; x_1^2 \mathrm{f}_i(x_1)\, x_2^2 \mathrm{f}_j(x_2)\, \frac{d^2\hat{\sigma}_{ij}}{dv dz}(v, z, p_{\mathrm{T}}, \mathcal{R}), \tag{3.19}$$

where p_{T} and y are the transverse momentum and rapidity of the signal jet, $V = 1 - e^{-y} p_{\mathrm{T}}/\sqrt{s}$, $VW = e^y p_{\mathrm{T}}/\sqrt{s}$ and $v = u/(t+u)$, with s, t, u the Mandelstam variables. Then $x_1 = VW/v/(1-z)$ and $x_2 = (1-V)/(1-v)/(1-z)$, and the partonic cross section factorizes as

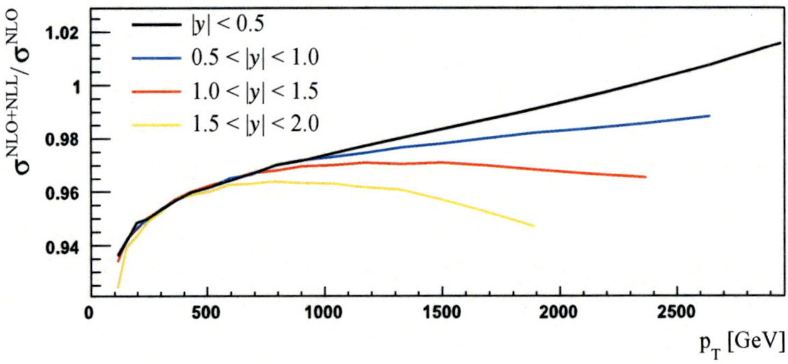

Fig. 3.9 The ratio of single inclusive jet production cross sections at NLO+NLL and NLO in pp collisions at $\sqrt{s} = 13\,\text{TeV}$, as a function of p_T in different regions of absolute rapidity $|y|$, for anti-k_T [64] jets with the distance parameter 0.7

$$
\begin{aligned}
\frac{d^2\hat{\sigma}_{ij}}{dvdz} = s \iiint ds_X ds_C ds_G \; & \delta(z_S - s_X - s_C - s_G) \\
& \times \Big(\operatorname*{Tr}_{\text{color}}[\mathcal{H}_{ij}(v, p_T, \mu_H, \mu_r)\mathcal{S}^G(s_G, \mu_{SG}, \mu_r)] \Big) \\
& \times \mathcal{J}_X(s_X, \mu_X, \mu_r) \\
& \times \left(\sum_k^{\text{splits}} \mathcal{J}_k(p_T\mathcal{R}, \mu_J, \mu_r) \otimes_\Omega \mathcal{S}_k^C(s_C\mathcal{R}, \mu_{SC}, \mu_r) \right).
\end{aligned} \tag{3.20}
$$

The $\mathcal{H}_{ij}$ are the hard functions for $2 \to 2$ scattering, available in Ref. [59]. The global soft function $\mathcal{S}^G$ accounts for radiation well outside the jet cones, and its form is given in Ref. [57]. The recoil, or inclusive, jet function $\mathcal{J}_X$ is extracted from Refs. [60, 61]. The signal jet functions $\mathcal{J}_k$ account for the energetic radiation inside the jet and are extracted at NLO from Ref. [62], whereas the soft radiation near the jet boundary is described by the collinear-soft (coft) functions $\mathcal{S}_k^C$. The form of the NLO bare coft function is given in Ref. [57]. The angular integrations $\otimes_\Omega$ are due to the mixing between the respective jet and coft functions [57]. The various functions are dependent on their characteristic invariant masses s, or p_T in the case of the signal jet function. They are evolved from their characteristic scales μ to a common renormalization scale μ_r by solving their respective RGEs, as in e.g. [63].

The threshold resummation results in an increase in the inclusive jet cross section, whereas $\mathcal{R}$-resummation leads to a decrease. In the low p_T end of the cross section, the $\mathcal{R}$-resummation dominates. The effects of resummation are illustrated in Fig. 3.9, showing the ratio of the inclusive jet production cross sections at NLO+NLL and NLO with the $p_T^{\max}$ scale choice for μ_r and μ_f. In general, the resummation is observed to decrease the cross section, although at low $|y|$ and high p_T the NLO+NLL cross section increases towards, and eventually slightly above, the NLO cross section.

References

1. Petrov AA, Blechman AE (2016) Effective field theories. World Scientific https://doi.org/10.1142/8619. https://www.worldscientific.com/doi/abs/10.1142/8619
2. Fermi E (1934) An attempt of a theory of beta radiation. 1 Z Phys 88:161–177
3. Becher T, Neubert M, Rothen L, Shao DY (2016) Effective field theory for jet processes. Phys Rev Lett 116:192001
4. Bauer CW, Fleming S, Luke ME (2000) Summing Sudakov logarithms in $B \to X_s\gamma$ in effective field theory. Phys Rev D 63:014006
5. Bauer CW, Fleming S, Pirjol D, Stewart IW (2001) An effective field theory for collinear and soft gluons: heavy to light decays. Phys Rev D 63:114020
6. Bauer CW, Stewart IW (2001) Invariant operators in collinear effective theory. Phys Lett B 516:134–142
7. Bauer CW, Pirjol D, Stewart IW (2002) Soft collinear factorization in effective field theory. Phys Rev D 65:054022
8. Beneke M, Chapovsky AP, Diehl M, Feldmann T (2002) Soft collinear effective theory and heavy to light currents beyond leading power. Nucl Phys B 643:431–476
9. Beneke M, Feldmann T (2003) Multipole expanded soft collinear effective theory with non Abelian gauge symmetry. Phys Lett B 553:267–276
10. Hill RJ, Neubert M (2003) Spectator interactions in soft collinear effective theory. Nucl Phys B 657:229–256
11. Georgi H (1990) An effective field theory for heavy quarks at low energies. Phys Lett B 240:447–450. ISSN: 0370-2693
12. Eichten E, Hill B (1990) An effective field theory for the calculation of matrix elements involving heavy quarks. Phys Lett B 234:511–516. ISSN: 0370-2693
13. Gao J (2013) CIJET: a program for computation of jet cross sections induced by quark contact inter actions at hadron colliders. Comput Phys Commun 184:2362
14. Gao J, Li CS, Yuan CP (2012) NLO QCD corrections to dijet production via quark contact interactions. JHEP 07:037
15. Weinberg S (1979) Phenomenological lagrangians. Phys A 96:327–340 (Deser S (ed))
16. Bertone G, Hooper D, Silk J (2005) Particle dark matter: evidence, candidates and constraints. Phys Rept 405:279–390
17. Goodman J et al (2010) Constraints on dark matter from colliders. Phys Rev D 82:116010
18. Petrov AA, Shepherd W (2014) Searching for dark matter at LHC with Mono-Higgs production. Phys Lett B 730:178–183
19. Jenkins EE, Manohar AV, Stoffer P (2018) Low-energy effective field theory below the electroweak scale: operators and matching. JHEP 03:016
20. Appelquist T, Carazzone J (1975) Infrared singularities and massive fields. Phys Rev D 11:2856–2861 (10 May 1975)
21. Weinberg S (1980) Effective gauge theories. Phys Lett B 91:51–55
22. Bilenky MS, Santamaria A (1994) One loop effective Lagrangian for a standard model with a heavy charged scalar singlet. Nucl Phys B 420:47–93
23. Schwartz MD (2014) Quantum field theory and the standard model. Cambridge University Press, Cambridge. ISBN: 978-1-107-03473-0, 978-1-107-03473-0
24. Weinberg S (1979) Baryon- and lepton-nonconserving processes. Phys Rev Lett 43:1566–1570 (21 Nov 1979)
25. Leung CN, Love ST, Rao S (1986) Low-energy manifestations of a new interactions scale: operator analysis. Zeitschrift für Phys C Part Fields 31:433–437. ISSN: 1431-5858
26. Buchmuller W, Wyler D (1986) Effective Lagrangian analysis of new interactions and flavour conservation. Nucl Phys B 268:621–653. ISSN: 0550-3213
27. Grzadkowski B, Iskrzynski M, Misiak M, Rosiek J (2010) Dimension-six terms in the standard model Lagrangian. JHEP 10:085
28. Giudice GF, Grojean C, Pomarol A, Rattazzi R (2007) The strongly-interacting light Higgs. JHEP 06:045

29. Brivio I, Trott M (2019) The standard model as an effective field theory. Phys Rept 793:1–98
30. Jenkins EE, Manohar AV, Trott M (2013) Renormalization group evolution of the standard model dimension six operators I: formalism and lambda dependence. JHEP 10:087
31. Jenkins EE, Manohar AV, Trott M (2014) Renormalization group evolution of the standard model dimension six operators II: Yukawa dependence. JHEP 01:035
32. Alonso R, Jenkins EE, Manohar AV, Trott M (2014) Renormalization group evolution of the standard model dimension six operators III: gauge coupling dependence and phenomenology. JHEP 04:159
33. Grojean C, Jenkins EE, Manohar AV, Trott M (2013) Renormalization group scaling of Higgs operators and $\Gamma(h->\gamma\gamma)$. JHEP 04:016
34. Alonso R, Chang H-M, Jenkins EE, Manohar AV, Shotwell B (2014) Renormalization group evolution of dimension-six baryon number violating operators. Phys Lett B 734:302–307
35. Lehman L (2014) Extending the standard model effective field theory with the complete set of dimension-7 operators. Phys Rev D 90:125023
36. Murphy CW (2020) Dimension-8 operators in the standard model effective field theory. JHEP 10:174
37. Terazawa H (1980) Sub quark model of leptons and quarks. Phys Rev D 22:184–199 (1 July 1980)
38. Eichten E, Lane KD, Peskin ME (1983) New tests for quark and lepton substructure. Phys Rev Lett 50:811 (Tran Thanh Van J (ed))
39. Eichten E, Hinchliffe I, Lane K, Quigg C (1984) Supercollider physics. Rev Mod Phys 56:579–707 (4 Oct 1984)
40. Collins PDB, Martin AD, Squires EJ (1989) Particle physics and cosmology
41. Fritzsch H, Mandelbaum G (1981) Weak interactions as manifestations of the substructure of leptons and quarks. Phys Lett B 102:319–322. ISSN: 0370-2693
42. Fritzsch H, Mandelbaum G (1982) The substructure of the weak bosons and the weak mixing angle. Phys Lett B 109:224–226. ISSN: 0370-2693
43. Harari H (1979) A schematic model of quarks and leptons. Phys Lett B 86:83–86. ISSN: 0370-2693
44. Shupe MA (1979) A composite model of leptons and quarks. Phys Lett B 86:87–92
45. Bars I (1982) All preon models within 10 constraints. Phys Lett B 114:118–124. ISSN: 0370-2693
46. Pati JC, Salam A (1974) Lepton number as the fourth color. Phys Rev D 10:275–289 [Erratum: Phys Rev D 11:703 (1975)]
47. Harari H, Mohapatra RN, Seiberg N (1982) Proton decays and neutron oscillations in the Rishon model. Nucl Phys B 209:174–182
48. Lane KD (1996) Electroweak and flavor dynamics at hadron colliders. arXiv: hep-ph/9605257
49. Tumasyan A et al (2022) Measurement and QCD analysis of double-differential inclusive jet cross sections in pp collisions at $\sqrt{s} = 13$ TeV. JHEP 02:142
50. Langacker P (2009) The physics of heavy Z$\prime$ gauge bosons. Rev Mod Phys 81:1199
51. Harris BW, Owens JF (2002) Two cutoff phase space slicing method. Phys Rev D 65:094032 (9 May 2002)
52. Catani S, Seymour MHA (1997) General algorithm for calculating jet cross-sections in NLO QCD. Nucl Phys B 485:291–419 [Erratum: Nucl Phys B 510:503–504 (1998)]
53. Buchalla G, Buras AJ, Lautenbacher ME (1996) Weak decays beyond leading logarithms. Rev Mod Phys 68:1125–1244 (4 Oct 1996)
54. Gao J, Li CS, Wang J, Zhu HX, Yuan C-P (2011) Next-to-leading QCD effect to the quark compositeness search at the LHC. Phys Rev Lett 106:142001
55. Chanowitz MS, Furman M, Hinchliffe I (1979) The axial current in dimensional regularization. Nucl Phys B 159:225–243
56. Becher T, Broggio A, Ferroglia A (2015) Introduction to soft-collinear effective theory (2015). https://doi.org/10.1007/978-3-319-14848-9. arXiv:1410.1892 [hep-ph]
57. Liu X, Moch S-O, Ringer F (2017) Threshold and jet radius joint resummation for single-inclusive jet production. Phys Rev Lett 119:212001

58. Liu X, Moch S-O, Ringer F (2018) Phenomenology of single-inclusive jet production with jet radius and threshold resummation. Phys Rev D 97:056026
59. Broggio A, Ferroglia A, Pecjak BD, Zhang Z (2014) NNLO hard functions in massless QCD. JHEP 12:005
60. Becher T, Neubert M (2006) Toward a NNLO calculation of the $\bar{B} \to X_s \gamma$ decay rate with a cut on photon energy. II: two-loop result for the jet function. Phys Lett B 637:251–259
61. Becher T, Bell G (2011) The gluon jet function at two-loop order. Phys Lett B 695:252–258
62. Liu X, Petriello F (2013) Resummation of jet-veto logarithms in hadronic processes containing jets. Phys Rev D 87:014018
63. Becher T, Neubert M (2006) Threshold resummation in momentum space from effective field theory. Phys Rev Lett 97:082001
64. Cacciari M, Salam GP, Soyez G (2008) The anti-kT jet clustering algorithm. JHEP 04:063

Chapter 4
LHC Measurements

This chapter presents a brief review of the Large Hadron Collider (LHC) [1], the main machine of the accelerator complex located at the European center of nuclear research (CERN) in Geneva, and the Compact Muon Solenoid (CMS) Experiment [2]. The LHC is discussed in Sect. 4.1, while the CMS Experiment is sketched in Sect. 4.2. The CMS measurements of inclusive jet and $t\bar{t}$ production cross sections interpreted in this thesis are introduced in Sects. 4.3 and 4.4, respectively.

4.1 The Large Hadron Collider

The LHC is a circular two-ring superconducting hadron accelerator placed in a 26.7 km long tunnel at CERN. The LHC is designed for accelerating and colliding protons (p) and lead (Pb) ions in two beams circulating the machine in opposite directions, in the pp, pPb and PbPb beam configurations.

During run I, from 2010 to 2012, the LHC operated at the center of mass energies of 2.76, 7 and 8 TeV for proton-proton collisions. After the first long shutdown in 2013–2015, the LHC started colliding protons at $\sqrt{s} = 13$ TeV for run II from 2015 to 2018, with occassional heavy-ion runs at 5.02 and 8.16 TeV. After the second long shutdown, run III has started in July 2022 at the designed center of mass energy of 13.6 TeV for pp collisions.

The protons are first extracted from ionizing hydrogen gas, and directed to the LINAC 2 linear accelerator for acceleration to 50 MeV. From there, they are injected into the proton synchrotron booster, where they reach 1.4 GeV. After that, they are accelerated to 25 GeV by the proton synchrotron (PS). The PS sends them to the Super Proton Synchrotron (SPS) for acceleration to 450 GeV. Finally, the protons are injected into LHC ring where they reach their final energy. For lead ions, the first stages of acceleration are the LINAC 3 linear accelerator and the low energy ion ring (LEIR), after which they follow the same route as the protons. A schematic picture of the CERN accelerator complex is given in Fig. 4.1.

T. Mäkelä, *Towards Global Interpretation of LHC Data*, Springer Theses,
https://doi.org/10.1007/978-3-031-29779-3_4

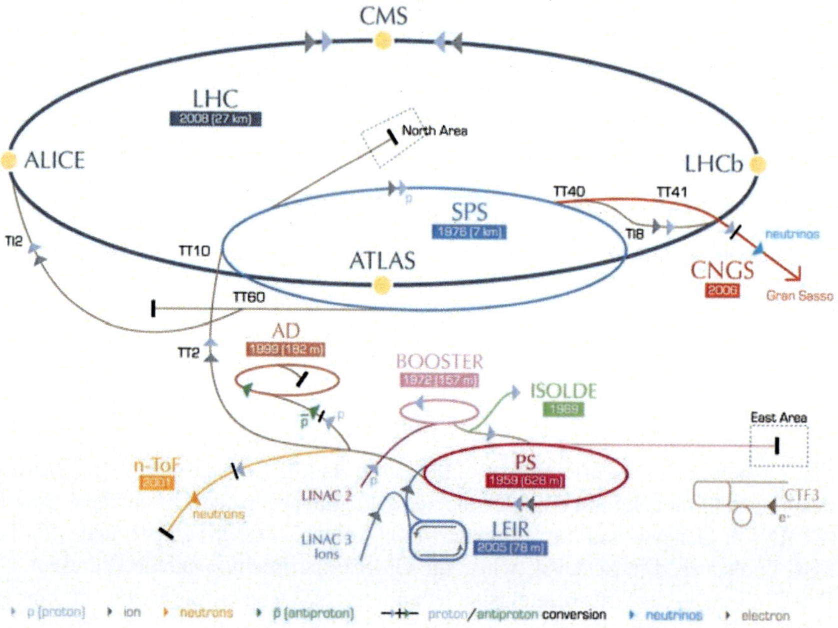

Fig. 4.1 The CERN accelerator complex. The acceleration stages are illustrated by denoting the fraction of the speed of light c at which the protons travel in the laboratory frame. Image taken from Ref. [3]

With the run II setup, a collision of two proton bunches occurred every 25 ns. The frequency of collision events in a collider is related to the *instantaneous luminosity*, which can be written in terms of the machine parameters as

$$\mathcal{L} = \frac{n_b N_b^2 f \gamma F}{4\pi \epsilon_n \beta^*}, \tag{4.1}$$

where n_b is the number of bunches, N_b is the number of particles per bunch, f the rotation frequency, γ the relativistic gamma factor, ϵ_n is the normalized transverse beam emittance, β^* is a function related to the transverse size of the beam at the collision point and F is the geometric luminosity reduction factor due to the crossing angle at the interaction point (IP). The value of $\mathcal{L}$ is determined via observables proportional to the number of pp collisions per bunch-crossing, e.g. the amount of clusters in an experiment's tracking detector. The number of events per second is given by $N_{\text{event}} = \mathcal{L}\sigma_{\text{event}}$, when the event has a cross section of σ_{event} [4]. Integrating the instantaneous luminosity over a time window then results in a measure for the amount of acquired data, called the *integrated luminosity*

$$\mathcal{L}_{\text{int}} \equiv \int dt \mathcal{L} = \frac{N}{\sigma} \tag{4.2}$$

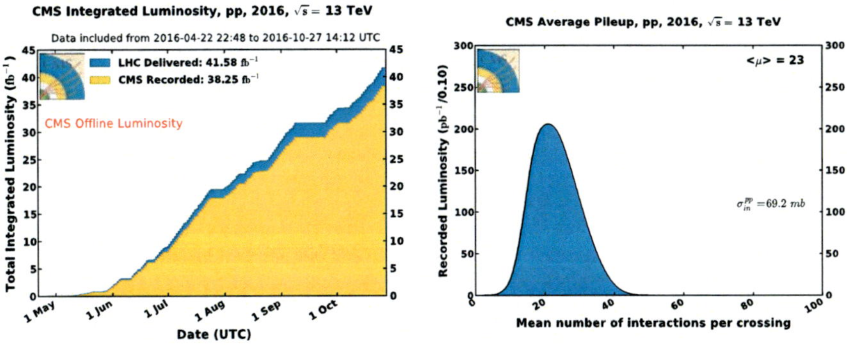

Fig. 4.2 *Left*: the total integrated luminosity as delivered by the LHC and as recorded by the CMS, using the best available offline measurement and calibrations. *Right*: the mean number of interactions per bunch crossing. The value assumed for the minimum bias cross section is the CMS recommended value of 69.2 mb, determined by the best agreement with data. Figures correspond to the 13 TeV pp run in 2016 and are taken from Ref. [6]

which has the units of an inverse cross section. The total integrated luminosity for pp collisions at $\sqrt{s} = 13$ TeV during 2016 is shown in Fig. 4.2, along with the mean number of pp interactions per bunch crossing in the same year. The presence of additional pp interactions besides the examined hard interaction, from the same or nearby bunch crossings, is referred to as *pileup*. The original design luminosity of the LHC is $1 \cdot 10^{34}\text{cm}^{-2}\text{s}^{-1}$ for pp collisions, and a peak instantaneous luminosity of $2 \cdot 10^{34}\text{cm}^{-2}\text{s}^{-1}$ was confirmed experimentally in 2018 [5]. The instantaneous luminosity is planned to increase from the original design value by a factor of five in the high-luminosity LHC upgrade, which is to be carried out after run III [5].

In general, an experimental measurement of a cross section is performed by counting the number of events for a given process and dividing by the $\mathcal{L}_{\text{int}}$ accumulated in the same time-interval. The cross section's kinematic dependence on an observable X, ergo the *differential cross section* is then given in terms of experimental quantities as

$$\frac{d\sigma}{dX} = \frac{N}{\varepsilon \mathcal{L}_{\text{int}} \Delta X} \tag{4.3}$$

where ε represents experimental efficiencies, and ΔX is the measured kinematic range (bin) [7].

There are four major experiments along the LHC ring: ALICE [8], ATLAS [9], CMS [2] and LHCb [10]. The two general-purpose experiments ATLAS and CMS are designed for studying SM processes and for performing searches for BSM physics, ALICE for heavy-ion collisions and LHCb for precision studies of CP violation and rare B-hadron decays. In this thesis, measurements at the LHC during run II, collected in 2016 by the Compact Muon Solenoid (CMS) experiment in proton-

proton collisions at $\sqrt{s} = 13\,\text{TeV}$, are used, corresponding to integrated luminosities of $33.5\,\text{fb}^{-1}$ for inclusive jet production, using jets reconstructed using the anti-k_T algorithm with the distance parameter 0.7, and $35.9 \pm 0.9\,\text{fb}^{-1}$ for $t\bar{t}$ pair production.

4.2 The Compact Muon Solenoid Experiment

The CMS Experiment is designed as a multi-purpose detector for performing ultimate precision tests of the SM, as well as searches for BSM physics [11]. It identifies stable particles produced in the collisions by measuring their momenta and energies, thereby recreating an image of the collision. The CMS detector is composed of several subdetector layers with different functionalities for obtaining information on the particles tracing the detector, as illustrated in Fig. 4.3.

A central feature of the CMS detector is the superconducting solenoidal magnet, which produces a homogeneous magnetic field of 3.8 T. The solenoid volume encompasses a silicon pixel and strip tracker, a lead-tungstate crystal electromagnetic calorimeter, and a brass and scintillator hadronic calorimeter, each composed of a barrel and two endcap sections. Several components of the muon system are embedded into the return yoke, and it is used as an additional hadron absorber.

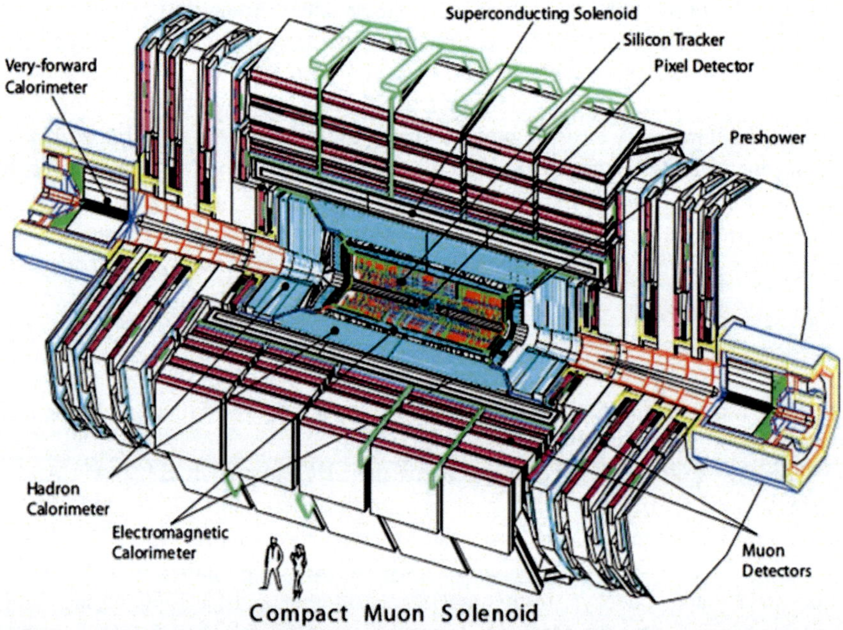

Fig. 4.3 An illustration of the components and layers of the CMS detector. Image taken from Ref. [2]

The coordinate system [12] of the detector is such that center is located at the nominal collision point inside the detector, the x-axis points towards the center of the LHC ring, the y-axis points upwards and the z-axis points towards the counter-clockwise direction (as viewed from above) of the beamline. The polar angle θ is defined with respect to the z-axis, while the azimuthal angle ϕ and the radial distance r are measured in the *transverse plane* xy, with ϕ starting from the x axis.

The kinematics of the produced objects are defined by the transverse momentum p_T and rapidity or the *pseudorapidity*

$$\eta = -\log\left(\tan\left(\frac{\theta}{2}\right)\right), \tag{4.4}$$

where theta is the polar angle.

4.2.1 The CMS Tracker

The tracking system is designed for measuring charged particle trajectories, and for reconstructing the primary vertices. With an impact parameter, defined as the transverse distance to the IP, resolution of the order of 100 μm, the reconstruction of secondary vertices from the decays of long-lived heavy particles is possible as well. The precise track information is used for the determination of particle momenta, positions and charges.

The CMS tracker is divided into an inner pixel detector and an outer strip detector, as illustrated in Fig. 4.4. The inner pixel detector consists of three cylindrical sensor layers mounted as close to the IP as possible, with the radial distances 4.4, 7.3 and 10.2 cm. A $r\phi$ resolution of 10 μm and a z resolution of approximately 20 μm are achieved with the 66 million pixels distributed around an area of 1 m^2.

The silicon strip detector covers the intermediate radial region of $20 < r < 120$ cm and an area of approximately 200 m^2. It is divided into the tracker inner barrel (TIB),

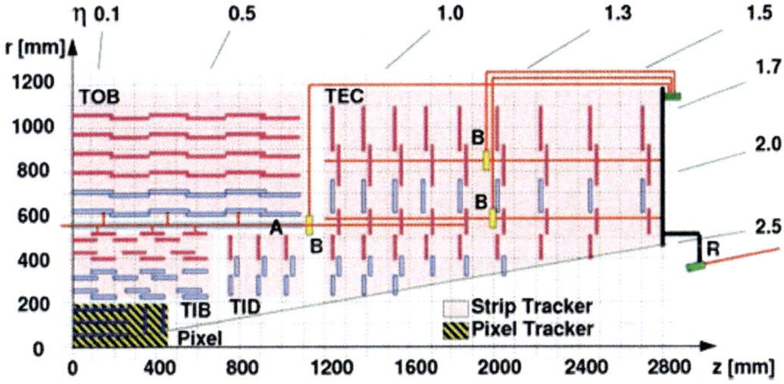

Fig. 4.4 A schematic of a quadrant of the tracker in the rz plane. Image taken from Ref. [13]

the tracker outer barrel (TOB), the tracker inner disks (TID) and the tracker endcap (TEC). Electron-hole pairs are produced as charged particles pass through the diode material, and they are collected by the silicon strips. In addition to the transverse plane measurements, the stereo modules provide a measurement in the rz plane.

Depending on the 20 different module geometries in the silicon-strip detector, the resolution in the $r\phi$ plane varies from 20 to 50 μm. In the z direction, the resolution is 500 μm. The CMS tracker covers an area extending up to $\eta \sim |2.5|$.

4.2.2 The Electromagnetic Calorimeter

The main purpose of the CMS electromagnetic calorimeter (ECAL) [14] is to measure the energy and position of photons, electrons and charged hadrons through the energy deposit of the produced showers. A schematic of the ECAL is given in Fig. 4.5. In combination with the tracker, it has an important role in the identification of particles. It consists of 76000 lead-tungstate ($PbWO_4$) crystals, which is a fast (80% of the light is emitted within 25 ns), dense and radiation-hard material [15].

The ECAL is divided into three subsystems: the electromagnetic barrel (EB) calorimeter, the electromagnetic endcap (EE) calorimeters and the preshower detector. The crystals in the EB calorimeter are mounted such that the axes of each crystal are tilted in both η and ϕ by 3° with respect to the direction to the IP. Due to this geometry, the particles do not pass along the boundary of adjacent crystals. The face cross section of a crystal is approximately 22×22 mm^2 at the inner side of the detector [12], and each crystal has a length of 230 mm, corresponding to 25.8 radiation lengths. The inner radius of the EB is 1.3 m, and it covers the region up to $|\eta| = 1.479$. The produced scintillation light is detected using avalanche photodiodes (APDs). The EE calorimeters are perpendicular to the beam line, covering

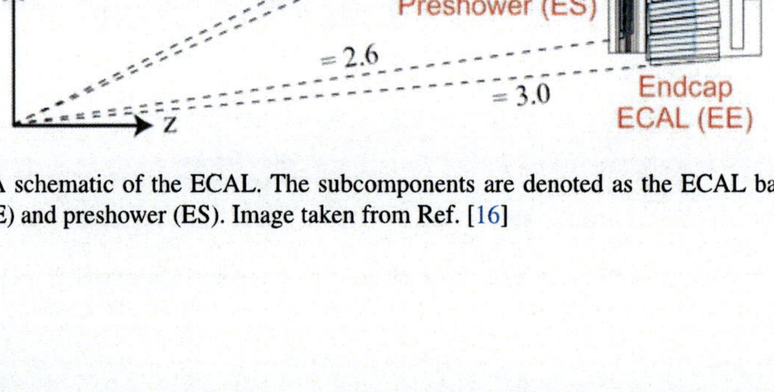

Fig. 4.5 A schematic of the ECAL. The subcomponents are denoted as the ECAL barrel (EB), endcap (EE) and preshower (ES). Image taken from Ref. [16]

a range of $1.479 < |\eta| < 3.0$. In the EE, the readout is performed using vacuum photo-triodes (VPT).

The preshower detector is a sampling calorimeter with fine granularity. It is installed within the fiducial region $1.65 < |\eta| < 2.61$, and serves mainly to improve the identification of $\pi^0 \to \gamma\gamma$ decays against prompt photons, as well as enhance the position determination of photons and electrons. The detector consists of alternatig layers of lead absorbers, initiating the electromagnetic showers, and planes of silicon strip detectors, placed after each absorber [12].

4.2.3 The Hadronic Calorimeter

The hadronic calorimeter (HCAL) [17] is a sampling calorimeter consisting of brass absorber plates and scintillator layers. It is used for measuring the energies of hadronic jets by absorbing their energy [15]. The CMS HCAL is divided into four regions, as shown in Fig. 4.6: the central barrel (HB), the outer barrel (HO), the endcap (HE) and the forward calorimeters (HF). The HB and HE completely enclose the ECAL, and are fully within the high magnetic field of the solenoid.

The HB covers up to $|\eta| = 1.4$, and consists of 36 wedges with alternating layers of 56.5 mm thick brass absorber plates and 3.7 mm thick plastic scintillator plates. The thickness increases from 5.82 interaction lengths in the central region to 10.6 at $|\eta| = 1.3$.

The HO is an array of scintillators placed outside the magnet within $|\eta| < 1.26$. It is used for identifying late starting showers, and improves the energy resolution for high p_T jets in the central region. It makes use of the solenoid as an absorber, and

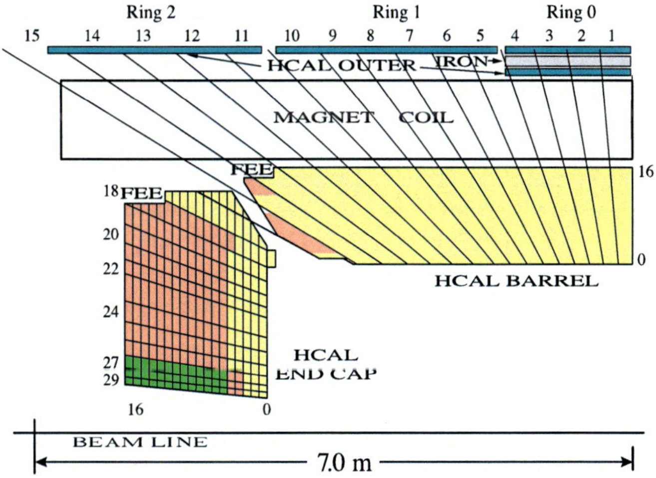

Fig. 4.6 The layout of the HCAL in the rz plane. Image taken from Ref. [18]

10 mm thick scintillator plates are placed before the muon system. However, due to the low absorber depth in the very central region, a HO scintillator layer is placed on both sides of a 19.5 cm thick sheet of iron.

The HE are in the range of $1.3 < |\eta| < 3.0$, and have larger absorber plates with a thickness of 97, and 9 mm gaps for the scintillator plates. Thus the absorber thickness is approximately 10 interaction lengths throughout the HE.

The extensivity of the energy measurement is ensured by the HF calorimeters at $3.0 < |\eta| < 5.0$, placed 11.2 m away from the IP, outside of the magnetic volume. They consists of 5 mm thick steel plates as the absorber, with embedded Cherenkov radiating quarz fibres as the sensitive material. This choice of design is based on the requirements on radiation hardness for the active medium in this region. The average energy deposit in this component is 760 GeV, in comparison to the 100 GeV expected for the rest of the detector. The HF is most sensitive to the electromagnetic component of hadronic showers, as it detects the Cherenkov radiation emitted by charged particles above the Cherenkov threshold, which is $E \geq 190$ keV for electrons.

The readout in HCAL is achieved by individual towers, with a cross section of $\Delta\eta \times \Delta\phi = 0.087 \times 0.087$ in the barrel, and 0.017×0.017 at high rapidities [19].

4.2.4 The Magnet and the Muon System

The large superconducting solenoid magnet in CMS delivers an axial and uniform magnetic field of 3.8 T within a length of 12.5 and 6 m free inner radius. It is large enough to host the tracker, ECAL and HCAL [19]. The high solenoidal magnetic field is necessary for good muon momentum resolution [15]. Outside the coil, the muon system is embedded in the iron yoke, which returns the magnetic flux and allows only muons and neutrinos to pass through.

The momenta and positions of muons produced at the IP are first measured in the tracking system. As the muons pass through the calorimeters, they deposit only ionization energy. After that, they reach the muon system consisting of several chambers embedded in the iron yoke. The muon system is designed for identifying and measuring minimum-ionizing muons.

The muon system is divided into three sub-systems with different detector technologies. Four layers of drift tube (DT) chambers cover the barrel region up to $|\eta| = 1.2$. At each side of the detector, 234 cathode strip chambers (CSCs) are installed, stretching up to $|\eta| = 2.4$. Resistive plate chambers (RPCs) are set up between the DTs and CSCs, reaching up to $|\eta| = 1.6$. The placement of the subsystems is illustrated in Fig. 4.7.

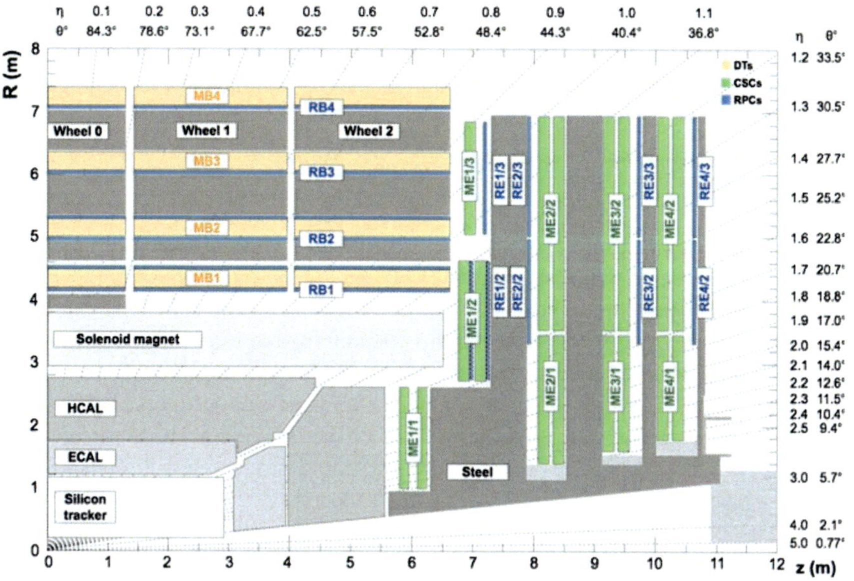

Fig. 4.7 The layout of one quadrant of the CMS detector. The DT stations are labelled as MB (muon barrel), and the CSCs as ME (muon endcap). The RPCs are mounted in both the barrel and the endcaps, where they are labelled RB and RE, respectively. Image taken from Ref. [20]

4.2.5 Trigger System

The bunch crossing rate of the LHC reaches 40 MHz at full capacity, translating to a pp collision rate of approximately 1 GHz [21]. Therefore, it is unfeasible to store all events, and a trigger system is required to reduce the rate to a 300 Hz. The selection of potentially interesting events is performed in two stages by the level-1 trigger (L1) and the high-level trigger (HLT) systems.

The L1 trigger performs a simplified event reconstruction via fast detector components, such as the calorimeters and the muon system, and a hardware processor based decision to accept or reject an event is made within 3.2 μs. The L1 selects events at a rate up to 100 kHz [22], based on candidate objects above pre-set thresholds that indicate the presence of physics objects such as photons, electrons, muons, or jets above given p_T thresholds. Also global event characteristics such as missing E_T are considered [12].

The events passing the L1 criteria are kept in buffers, and forwarded to the HLT, which is a computer farm of $\sim$ 13000 CPUs. The HLT performs further event selection with more complex offline-quality reconstruction algorithms. Several processing steps are run on the data in a predefined order, called the HLT-path. The physics objects are reconstructed, and selections are made based on the objects. The complexity of the algorithms and reconstruction refinement are increased in each step.

For instance, the data from the calorimeters and muon system is processed before the more sophisticated track reconstruction.

4.2.6 The Identification of Particles

As the particles emerging from the collisions interact with the detector matter, they leave different signatures depending on the type of the particle, which are then identified by the subdetectors. This is illustrated in Fig. 4.8. The pp collisions give rise to interaction vertices, spreading around the nominal interaction point along the beam axis The trajectories and vertices of charged particles entering the tracking system are reconstructed, and measurements of the position and the four-momentum are performed. Energy deposits in neighboring cells of the ECAL and HCAL are detected as clusters. The muon detectors outside the solenoid produce hits in the additional tracking layers to capture minimum-ionizing muons [19].

Hadronic jets are the most common objects in high-energy collider experiments. These are seen in the detectors as sprays of charged and neutral hadrons, leptons and photons. The energies of jets are measured from the hadronic and electromagnetic showers deposited in the HCAL and ECAL. Photons are identified from ECAL clusters with no associated track, while electrons are identified by a track in the inner

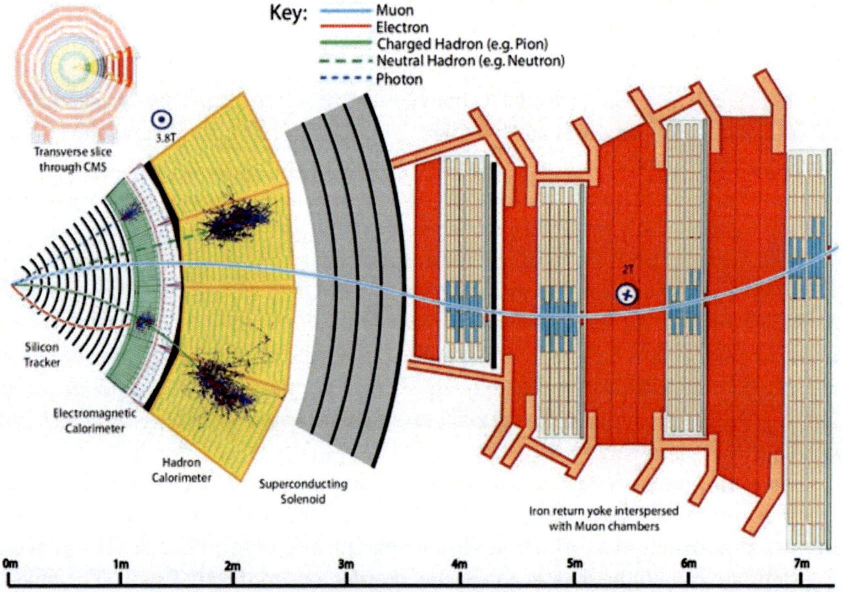

Fig. 4.8 The signatures of various particles traveling through the CMS detector, starting from the interaction point on the left of this azimuthal cross section schematic. Image taken from Ref. [19]

tracking system and a cluster in ECAL. The track and the cluster must match, and they should not be connected to the HCAL. Finally, muons are identified by the hits in the muon detectors and inner tracker.

A global event description is provided by the particle-flow (PF) algorithm [19], including the identification of all particles produced in a collision, and their discrimination against particles produced in pileup collisions. The PF simplifies particle reconstruction by correlating different detector components, and produces physics objects with a precise four-momentum and high resolution. Via the use of an iterative tracking algorithm, the tracker hits are used for building tracks while keeping the fake rate small. The tracks are associated with vertices, and the vertex with the largest summed physics-object p_{T}^2 value is defined as the *primary vertex* (PV). The *PF clusters* are formed by using cluster seeds, calorimetry cells with an energy that is above a given threshold and larger than the energy of the neighboring cells [19]. The combinations of PF tracks and PF clusters are called PF Blocks. Based on the PF Blocks, the algorithm creates PF Candidates. PF reconstruction is also used for the reconstruction of the missing transverse momentum, $\vec{\not{E}}_T$. It is defined as the negative vectorial sum of the transverse momenta of the PF particles. Particles creating $\vec{\not{E}}_T$ are considered to be neutrinos or weakly interacting particles not captured by the detector.

4.2.7 Jet Reconstruction

In the CMS double-differential inclusive jet cross section measurement [23], the jets are reconstructed offline from PF objects using the anti-k_{T} algorithm [24], run in the laboratory frame with distance parameters of 0.4 and 0.7, as implemented in the FastJet [25] package.

A typical jet event produced in a pp collision leaves the signature of energy deposits in both ECAL and HCAL, pointing to the nominal interaction vertex. The tracks are then associated to the calorimeter clusters by the PF algorithm, and charged hadrons are identified by a successful association of tracks to the clusters in the calorimeters. The tracks that are associated, are removed from the algorithm's input list. After all tracks are processed, the remaining clusters in ECAL are considered photons, and those in HCAL are labeled as neutral hadrons. Additionally, it is possible to identify muons and electrons inside the jet.

Once all particles are identified and reconstructed, they are clustered with jet algorithms such as the anti-k_{T} algorithm. The resolution on the momenta and positions of the PF jets is vastly improved by the tracking detector and the high granularity of the ECAL, through independent measurements of charged hadrons and photons within a jet. Together, these particles constitute nearly 85% of the energy of a jet [26].

The momentum of a jet is computed as the vector sum of all PF candidate momenta in the jet. This is determined from simulation to be on average within 5–10% of the true momentum, over the whole p_{T} spectrum and detector acceptance. The pileup can contribute additional tracks and energy deposits to the calorimeters, and thus increase the detector-level jet momentum. To mitigate this effect, tracks identified as originating from pileup vertices are discarded, and an offset correction is applied to correct for the remaining contributions [23].

The measured jet energies are corrected to the true energy by *jet energy scale* (JES) corrections, which are applied in the form of a global multiplicative factor to the jet four-momenta. Jet energy corrections are derived from simulation studies so that the average measured response of jets becomes identical to that after applying the corrections for detector effects, which is referred to as the *particle level*. The JES corrections include the *offset correction*, removing everything not related to the pp collision of the hard interaction, such as pileup and electronic noise. Next, the *MC calibration* corrects for the main non-uniformities in pseudorapidity and non-linearities in p_{T}, as calorimeters typically have a non-linear response, and the correction is identical for data and MC. Finally, the *residual corrections* account for finer corrections between the data and MC, and are only applied to the data; in situ momentum balance measurements in dijet, photon+jet, Z+jet, and multijet events are used for determining any remaining differences between the JES in data and simulation [26]. Further selection criteria are applied to each jet, so that jets potentially affected by instrumental effects or reconstruction failures are removed [27].

4.3 The CMS Measurement of Double-Differential Inclusive Jet Production Cross Section at 13 TeV

In this thesis, the measurements of inclusive jet production in pp collisions at the LHC at $\sqrt{s} = 13$ TeV are interpreted in terms of QCD and SMEFT. While the measurement itself is not the topic of this work, a large part of the investigation of this thesis is dedicated to the study of the systematic uncertainties in the inclusive jet cross sections and their correlations. The data were collected with the CMS experiment in 2016, and correspond to an integrated luminosity of 36.3 fb^{-1} and 33.5 fb^{-1} for jets reconstructed with the distance parameter $\mathcal{R} = 0.4$ and $\mathcal{R} = 0.7$, respectively [28]. The difference in the integrated luminosities is explained by the activation of the triggers for the $\mathcal{R} = 0.7$ jets after 2.8 fb^{-1} of data had been taken. In the QCD analysis which is the main topic of the thesis, the measurement using $\mathcal{R} = 0.7$ jets is used. In the following, a brief description of the measurement is given.

The measurement is performed double-differentially in bins of rapidity y and p_{T}, and the cross section reads

$$\frac{d^2\sigma}{dp_{\mathrm{T}}dy} = \frac{N_{\mathrm{jets}}^{\mathrm{eff}}}{\mathcal{L}\Delta p_{\mathrm{T}}\Delta y}, \tag{4.5}$$

where $N_{\mathrm{jets}}^{\mathrm{eff}}$ is the effective number of jets per bin after the corrections for detector effects have been applied. $\mathcal{L}$ is the integrated luminosity, while Δp_{T} and Δy are the bin widths in the jet transverse momentum and rapidity, respectively [23].

The primary vertex is required to be within $|z_{\mathrm{PV}}| < 24$ cm longitudinal and $\rho_{\mathrm{PV}} < 2$ cm radial distance from the nominal interaction point. To remove the effect of detector noise, the jets must fulfill quality criteria based on their constituents. Corresponding to the tracker acceptance, the jets must be reconstructed within $|y| < 2.5$. Jets reconstructed in detector regions that correspond to defective zones in the calorimeters are excluded, but recovered later in the unfolding procedure. The effect is uniform as a function of p_{T}, and of the order of a percent [23].

Various reconstruction failures, malfunctions in the detector, or noncollision backgrounds may lead to anomalous high-$p_{\mathrm{T}}^{\mathrm{miss}}$ events. These are rejected by event filters designed to identify more than 85–90% of ungenuine high-$p_{\mathrm{T}}^{\mathrm{miss}}$ events, with a mistagging rate below 0.1%. Due to the high jet rates in pp collisions, only every nth triggered event is actually stored, where n is called the *prescale factor*. The measurement makes use of prescaled single-jet triggers, requiring at least one jet with jet $p_{\mathrm{T}}^{\mathrm{HLT}}$ above a given threshold in the event. All triggers except the one with the highest threshold are prescaled, and correspond to different effective integrated luminosities [23].

A gradual shift in the timing of the inputs of the ECAL first-level trigger in the region $|\eta| > 2.0$, referred to as *prefiring*, caused a specific trigger inefficiency during the data taking in 2016. For events containing a jet with p_{T} larger than ≈ 100 GeV in the region $2.5 < |\eta| < 3.0$, the efficiency loss is approximately 10–20%, depending on p_{T}, η, and the data taking period. Correction factors were computed from data and applied to the acceptance evaluated by simulation.

In the measurement, the jets are weighted event by event. This is in contrast to the previous measurements [29–32] by the CMS Collaboration, where each trigger's contribution is normalized with respect to its effective luminosity. Due to the nonlinearity of the trigger rate as a function of the instantaneous luminosity and the JES corrections varying with time and pileup conditions, the choice of weighting method makes a difference particularly in terms of the spectrum's smoothness, even after unfolding [23]. The JES corrections are applied according to the standard CMS procedure [26]. To ensure and preserve the smoothness of the cross sections, an additional smoothing procedure is applied to the JES corrections, which are originally parametrized with linear splines in bins of p_{T} [33].

The detector response is modelled with GEANT4 [34], and the pileup is simulated with PYTHIA 8 [35] with the CUETP8M1 tune [36] corresponding to the pileup conditions in the data. Different detector effects on the measurement of a differential distribution are usually corrected for by the *unfolding* procedure. This includes effects such as the limited resolution, the acceptance, pile-up, inhomogeneities and anisotropies. Such effects may lead to smearing due to migrations, reconstruction inefficiencies caused by missing entries, or background from fake entries. The event count related to quantity generated in bin i and reconstructed in bin j is stored into a *response matrix*. Normalizing the response matrix row by row yields the *probability matrix* **A**, with the component A_{ij} corresponding to the probability that a quantity is measured in bin j when its true value is in bin i. The probability matrix is shown for jets reconstructed with the distance parameter 0.7 in Fig. 4.9, which shows five

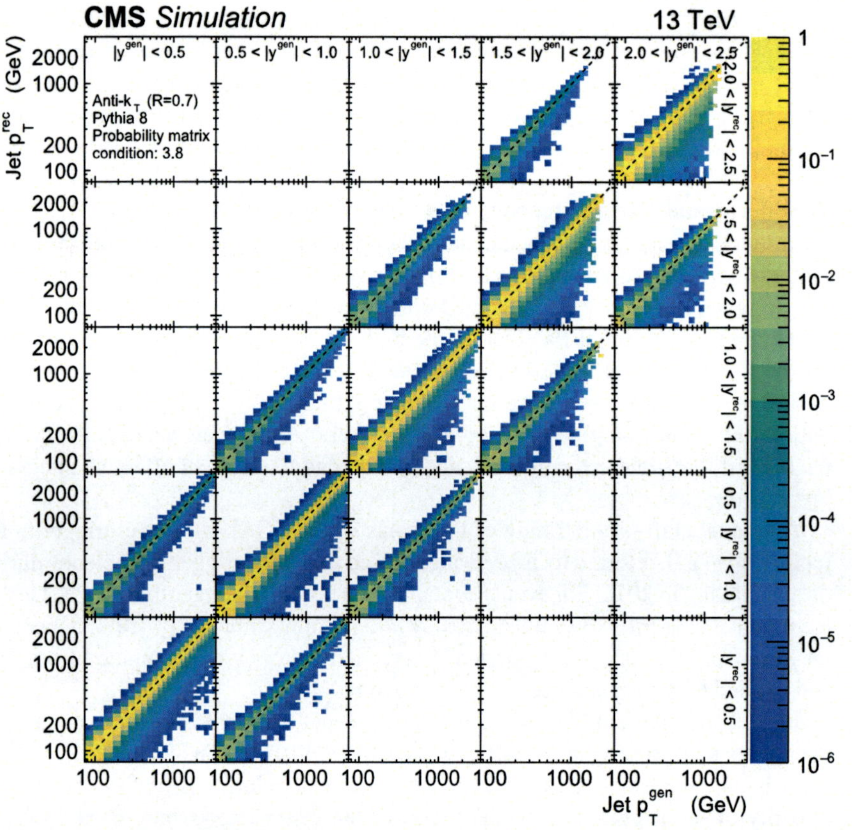

Fig. 4.9 The probability matrix, estimated with a simulated sample based on PYTHIA 8. The jets are clustered using the anti-k_T algorithm with $\mathcal{R} = 0.7$. The horizontal and vertical axes correspond to jets at the particle and detector levels, respectively. The 5×5 blocks correspond to jet rapidity bins as indicated by the labels in the top row and rightmost column, while the horizontal and vertical axes of each block correspond to the jet transverse momentum. The rows are normalized to unity, and the color range covers a range from 10^{-6} to 1, indicating the probability that a particle level jet generated with p_T^{gen} and $|y|^{gen}$ is reconstructed at detector level with values of p_T^{rec} and $|y|^{rec}$. Migrations outside of the phase space are not included, and migrations across y bins only occur among adjacent rapidity bins. The dashed lines indicate the diagonal bins in each rapidity cell [23]

rapidity bins. However, the $2.0 < |y| < 2.5$ bin was only kept in the analysis until the unfolding, and excluded after that.

The unfolding is performed two-dimensionally using least-square minimization. A truth vector $\mathbf{x}$ at hadron level is then determined by minimizing the function

$$\chi^2 = \min_{\mathbf{x}} \left[(\mathbf{A}\mathbf{x} + \mathbf{b} - \mathbf{y})^{\mathrm{T}} \mathbf{V}^{-1} (\mathbf{A}\mathbf{x} + \mathbf{b} - \mathbf{y}) \right], \tag{4.6}$$

where **b** is the background vector at detector level and **y** is the measurement vector, with **b** and **y** at the detector level, while **V** is the covariance matrix of the detector-level data. **V** describes the statistical uncertainties associated with the data, including correlations, and with the backgrounds. The solution to Eq. (4.6) reads

$$\mathbf{x} = (\mathbf{A}^T\mathbf{V}^{-1}\mathbf{A})^{-1}\mathbf{A}^T\mathbf{V}^{-1}(\mathbf{y} - \mathbf{b}). \tag{4.7}$$

The dominant bin-to-bin fully correlated uncertainties in the measurement are the JES uncertainties. Variations of the JES and prefiring corrections are applied to the data at the detector level, and mapped to the particle level by repeating the unfolding. The systematic effects related to the JER and pileup profile correction are varied in the simulation sample. These are propagated to the particle level by repeating the unfolding procedure. The normalization of the estimates of the inefficiencies and backgrounds, obtained from MC simulation, are varied separately within a conservative estimate of 5%. This covers a potential model dependence in phase space migrations and the impact of the matching algorithm in the unfolding procedure.

To assess the model dependence in the unfolding for p_T and $|y|$, a model uncertainty is derived as the difference between the nominal cross section obtained using a PYTHIA 8 simulation, and a modified simulation where the inclusive jet spectrum is corrected to match the data. The modified simulation is obtained via a smooth correction that is a function of p_T, y, and jet multiplicity. The effect is most prominent in the $|y| > 1.5$ region, where the data are not well described by PYTHIA 8. A MADGRAPH5_aMC@NLO [37, 38] simulation is used as an alternative to PYTHIA 8, and the results of both simulations are in good agreement. The relative precision of 1.2% in the integrated luminosity calibration [28] is applied as a fully correlated uncertainty. To remove the bin-to-bin fluctuations in the systematic variations, a smoothing procedure based on Chebyshev polynomials is applied following the method of [33].

Finally, all the uncorrelated and partly correlated uncertainties among p_T and y bins need to be considered. The dominant contribution to the uncorrelated uncertainties is due to the inclusive jet measurement being based on multiple jets recorded in each event. To account for differences between alternative methods to determine the trigger efficiency, an additional uncorrelated systematic uncertainty of 0.2% is added before the unfolding. Furthermore, the simulated distributions contain statistical fluctuations. The different sources are no longer distinguishable after unfolding. Furthermore, anti-correlations among directly neighbouring bins are introduced by the unfolding procedure.

The contributions of various uncertainties are shown in Fig. 4.10, where the color bands indicate bin-to-bin fully correlated uncertainties, while the vertical error bars indicate the uncorrelated uncertainties. The correlation matrix after the unfolding is shown in Fig. 4.11.

The resulting unfolded distribution is compared to the QCD prediction at NNLO in Fig. 4.12 [23], showing the ratios of the measured cross sections to the NNLO QCD predictions using different choices for the scales μ_r and μ_f. The comparison of the resulting unfolded distribution to the QCD prediction at NLO+NLL is given in Fig. 4.13, where the NLO+NLL predictions are shown using various PDFs sets.

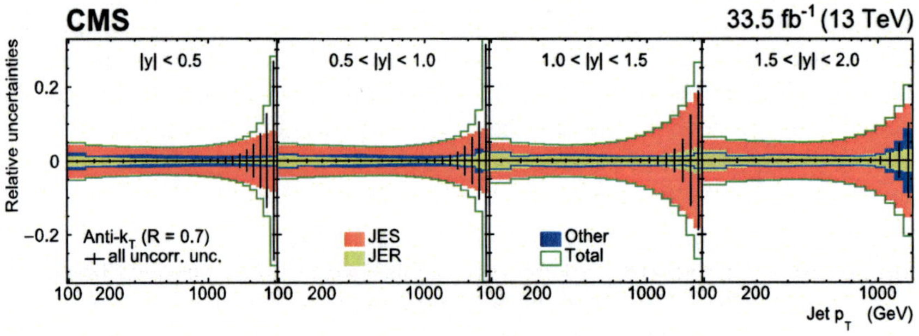

Fig. 4.10 Relative uncertainties in the double-differential cross section, as functions of jet transverse momentum (horizontal axis) and rapidity (cells), for jets clustered using the anti-k_T algorithm with $\mathcal{R} = 0.7$. The systematic uncertainties are shown in different noncumulative bands: JES uncertainties (red), JER uncertainties (yellow), and all other sources (blue), including the integrated luminosity uncertainty, the model uncertainty, uncertainties in the migrations in and out of the phase space, as well as uncertainties in various inefficiencies and backgrounds. The error bars include the statistical uncertainties from data, and from the PYTHIA 8 sample used in unfolding, and the binwise systematic uncertainties, all summed in quadrature. The total uncertainty (green) includes all systematic and statistical uncertainties summed in quadrature [23]

At high p_T, the predicted cross sections are in general smaller (higher) than the experimentally measured values at low (high) rapidities. The theoretical uncertainties are larger at high p_T and are dominated by the PDF uncertainties. Notably, the scale uncertainties are significantly smaller at NNLO than at NLO+NLL. The predictions with μ_r and μ_f set to jet p_T result in a harder p_T spectrum than the case with the H_T parton scale choice. All predictions describe the data well within the experimental and theory uncertainties.

4.4 The CMS Measurements of Differential $t\bar{t}$ Production Cross Sections at 13 TeV

The extraction of the top quark MSR mass presented in this thesis utilizes the single-differential $t\bar{t}$ production cross section data of Ref. [39], and the NLO SMEFT and QCD analyses performed in the context of this thesis feature the normalized triple-differential $t\bar{t}$ production cross section data of Ref. [40] in addition to the inclusive jet production cross section. The $t\bar{t}$ production cross section data are recorded by the CMS Collaboration in pp collisions at $\sqrt{s} = 13\,\text{TeV}$ at the LHC in 2016, corresponding to an integrated luminosity of $35.9\,\text{fb}^{-1}$.

The signal is defined by events in which both t quarks decay into a b quark and a W boson, and each W decays leptonically. All other $t\bar{t}$ events are considered as background [40]. The signal consists of three distinct channels: two same-flavor channels, with either two electrons or two muons in the final state, and the different-

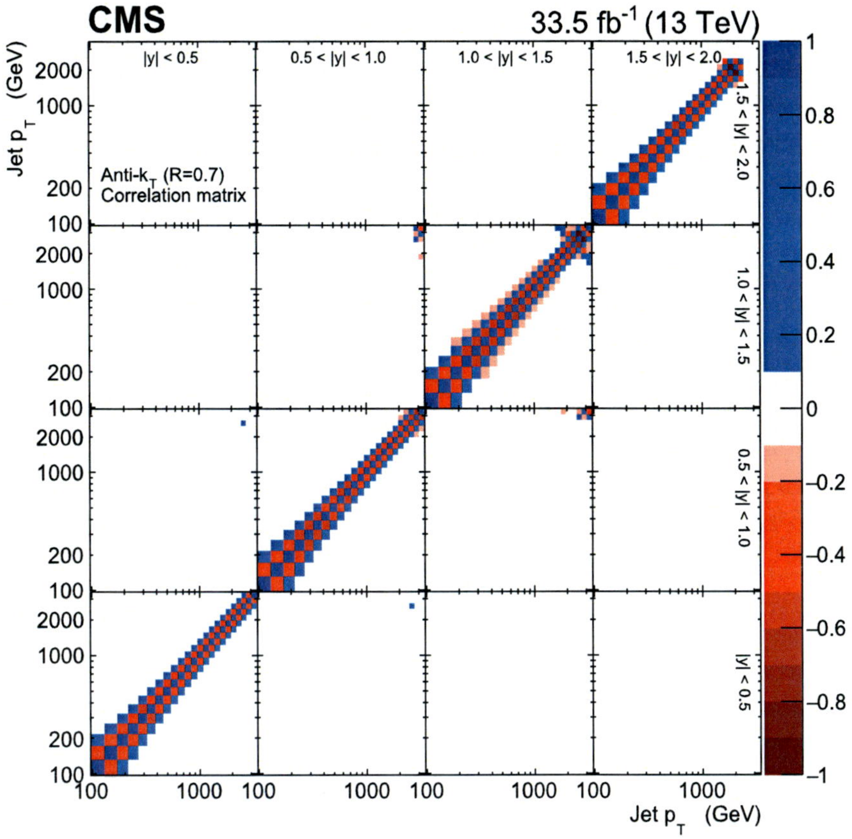

Fig. 4.11 The correlation matrix at the particle level, for jets clustered using the anti-k_T algorithm with $\mathcal{R} = 0.7$, containing contributions from the data and from the PYTHIA 8 sample used for performing the unfolding. The 4×4 blocks correspond to the jet rapidity bins as indicated by the labels in the top row and rightmost column, while the horizontal and vertical axes in each block correspond to the jet p_T [23]

flavor channel with one electron and one muon. For the final result, the three channels are combined [40].

The events are required to contain at least one electron and one muon with opposite electric charges, and a minimum of two jets, of which at least one is b tagged [40, 41]. The jets are reconstructed by combining PF candidates using the anti-k_T algorithm with the distance parameter $\mathcal{R} = 0.4$, and jets with $p_T > 30$ GeV and $|\eta| < 2.4$ are considered. Detector response effects are corrected for by applying p_T and η-dependent jet energy corrections [26]. Further details on the triggers and event selection are given in Refs. [39–41].

The invariant mass of the $t\bar{t}$ system is estimated by a kinematic reconstruction algorithm as described in Ref. [42], examining all possible combinations of recon-

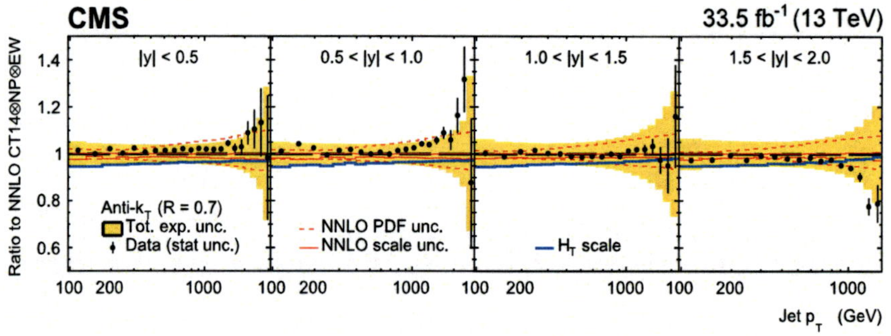

Fig. 4.12 The double-differential cross section of inclusive jet production, as a function of p_T and $|y|$, for jets clustered using the anti-k_T algorithm with $\mathcal{R} = 0.7$, presented as ratios to the QCD predictions corrected for NP and EW effects, using CT14nnlo PDF and with the QCD scales set to jet p_T and, alternatively, H_T (blue solid line) [23]

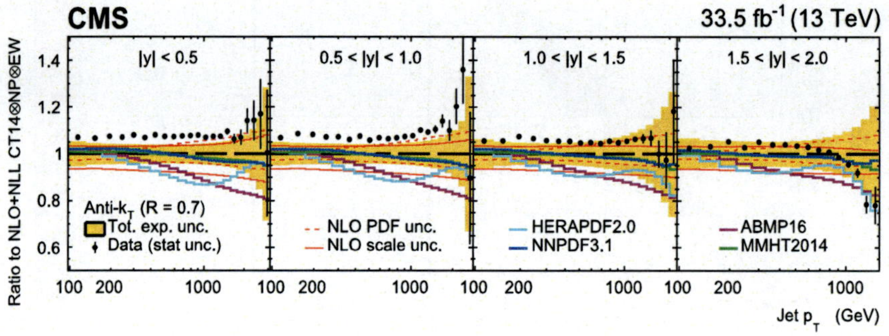

Fig. 4.13 The double-differential cross section of inclusive jet production, as a function of p_T and $|y|$, for jets clustered using the anti-k_T algorithm with $\mathcal{R} = 0.7$, presented as ratios to the QCD predictions corrected for NP and EW effects, using CT14nlo PDF [23]

structed jets and leptons. The algorithm solves a system of equations assuming that the reconstructed W boson has an invariant mass of 80.4 GeV and that the only contributions to the missing transverse momentum arise from the two neutrinos from the leptonic decays of the W bosons.

The sources of systematic uncertainty are divided into experimental and modelling uncertainties. The experimental uncertainties are related to the corrections to the MC simulation, including uncertainties associated with lepton identification and trigger efficiencies, jet energy scale [26] and resolution [27], lepton energy scales, b tagging efficiencies [43], and the uncertainty in the integrated luminosity [44]. The modelling uncertainties are related to the simulation of the $t\bar{t}$ signal, including scale variations for the matrix-element in POWHEG simulation [45, 46] and for the parton shower [47], variations in the matching scale between the matrix element and the parton shower [48], uncertainties in the underlying event tune [48], the parton distributions [49], the branching fraction and fragmentation function of B hadrons [50,

51], and uncertainties related to the choice of model for color reconnection [52, 53]. Following Refs. [41, 54, 55], an uncertainty accounting for the observed difference in the top quark p_T distribution shape between data and simulation [42, 56, 57] is applied. The dependence on the top quark width is found negligible [39]. Other sources of uncertainty include the modelling of additional pp interactions within the same or nearby bunch crossings, as well as the normalization of background processes. The normalization of each background process is assigned an uncertainty of 30%. Further details of the uncertainties are given in Ref. [58].

The single-differential cross section is measured in four bins of the invariant mass of the $t\bar{t}$ system: $m_{t\bar{t}} < 420$ GeV, $m_{t\bar{t}} \in [420, 550]$ GeV, $m_{t\bar{t}} \in [550, 810]$ GeV and $m_{t\bar{t}} > 810$ GeV. The cross section is determined at the parton level by performing a maximum-likelihood fit to distributions of final state observables, and treating the systematic uncertainties as nuisance parameters, as detailed in Ref. [39]. However, the kinematic reconstruction algorithm requires an assumed value m_t^{kin} for the top quark mass. In the determination of the single-differential $t\bar{t}$ production cross section of Ref. [39], any possible bias due to the choice of this value is avoided by including the dependence on m_t^{kin} in the fit. The dependence is estimated by repeating the kinematic reconstruction and event selection with m_t^{kin} set to three different values: 169.5, 172.5, and 175.5 GeV. The top quark mass m_t^{MC} used in the MC simulation is varied accordingly, and the parameter $m_t^{kin} = m_t^{MC}$ is treated as a free parameter in the fit [39].

The triple-differential $t\bar{t}$ production cross section is measured as a function of the invariant mass $m_{t\bar{t}}$ and rapidity $y_{t\bar{t}}$ of the $t\bar{t}$ system, in categories of additional jets N_{jet} produced in association with the $t\bar{t}$ pair. Since the measurements are defined at parton level, they are corrected for hadronization, detector resolution and inefficiency effects. A regularized unfolding process is performed simultaneously in bins of the three variables in which the cross sections are measured. The normalized differential $t\bar{t}$ cross section is obtained by dividing the distribution by the measured total inclusive cross section of $t\bar{t}$ production, which is evaluated as an integral over all bins in the respective observables [40].

In the triple-differential $t\bar{t}$ production cross section measurement, the issue with the dependence of the kinematic reconstruction on m_t^{kin} is resolved by applying the *loose kinematic reconstruction* algorithm. In this algorithm the $t\bar{t}$ kinematic variables are reconstructed without the top quark mass constraint, via reconstructing the $\nu\bar{\nu}$ system instead of the individual ν and $\bar{\nu}$. All the lepton and jet combinations in the event with an invariant mass $M_{\ell b} < 180$ GeV are considered, and the combinations are graded based on the presence of b-tagged jets. If a combination contains equally many b-tagged jets, the highest-p_T jets are selected. The transverse momentum vector of the $\nu\bar{\nu}$ system is set equal to $\mathbf{p}_T^{miss}$, whereas its longitudinal momentum and energy are set accordingly with those of the charged lepton pair. Additionally, the invariant mass of the neutrino pair is constrained to positive values, and the invariant mass of the W bosons is required to be above or equal to twice the mass of a single W boson.

The measurements of the cross section as a function of $m_{t\bar{t}}$, ranging from the threshold to the TeV region, are highly sensitive to the top quark mass, and the single-differential $t\bar{t}$ production cross section data are used in this thesis to extract the top quark mass in the MSR scheme. Moreover, the production of $t\bar{t}$ in association with extra jets provides additional sensitivity to m_t^{pole} due to gluon radiation depending on m_t^{pole} via threshold and cone effects [59], and can be used for constraining α_S at the scale of the top quark mass. Altogether, the triple-differential $t\bar{t}$ production cross section measurement is used in the SMEFT and QCD analyses at NLO, performed in the context of this thesis, to constrain the gluon PDF, α_S and m_t^{pole} [40].

References

1. Bruning OS et al (2004) LHC design report. CERN, Geneva. https://doi.org/10.5170/CERN-2004-003-V-1. https://cds.cern.ch/record/782076
2. Collaboration CMS (2008) The CMS experiment at the CERN LHC. JINST 3:S08004
3. Lefevre C (2008) The CERN accelerator complex. https://cds.cern.ch/record/1260465
4. Evans L, Bryant P (2008) LHC machine. J Instrum 3:S08001–S08001 Aug.
5. Zurbano Fernandez I et al (2020) High-luminosity large hadron collider (HL-LHC): technical de sign report 10 (Béjar Alonso I et al (eds)). https://doi.org/10.23731/CYRM-2020-0010
6. Collaboration C (2022) CMS luminosity—public results. Accessed 12 Feb 2022. https://twiki.cern.ch/twiki/bin/view/CMSPublic/LumiPublicResults
7. Rabbertz K (2017) Jet physics at the LHC. ISBN: 978-3-319-42113-1. https://doi.org/10.1007/978-3-319-42115-5
8. Collaboration A (2008) The ALICE experiment at the CERN LHC. JINST 3:S08002–S08002 Aug
9. Collaboration A (2008) The ATLAS experiment at the CERN large hadron collider. JINST 3:S08003–S08003 Aug.
10. Collaboration L (2008) The LHCb detector at the LHC. JINST 3:S08005–S08005 Aug.
11. Bayatian GL et al (2007) CMS physics: technical design report volume 2: physics performance. J Phys G 34. Revised version submitted on 22 Sept 2006 17:44:47, 995–1579. 669 p
12. Bayatian GL et al (2006) CMS physics: technical design report volume 1: detector performance and software. CERN, Geneva. https://cds.cern.ch/record/922757
13. Chatrchyan S et al (2014) Alignment of the CMS tracker with LHC and cosmic ray data. JINST 9:P06009
14. The CMS electromagnetic calorimeter project: technical design report. CERN, Geneva (1997). https://cds.cern.ch/record/349375
15. Green D (2010) At the leading edge: the ATLAS and CMS LHC experiments. World Scientific. ISBN: 9789814304672
16. Benaglia A (2014) The CMS ECAL performance with examples. JINST 9:C02008 (Govoni P et al (eds))
17. The CMS hadron calorimeter project: technical design report. CERN, Geneva (1997). https://cds.cern.ch/record/357153
18. Chatrchyan S et al (2009) Performance of CMS hadron calorimeter timing and synchronization using test beam, cosmic ray, and LHC beam data. JINST 5:T03013, 31 p
19. Sirunyan AM et al (2017) Particle-flow reconstruction and global event description with the CMS detector 12:P10003–P10003
20. Sirunyan AM et al (2018) Performance of the CMS muon detector and muon reconstruction with proton-proton collisions at $\sqrt{s}$ = 13 TeV. JINST 13:P06015
21. Khachatryan V et al (2017) The CMS trigger system. JINST 12:P01020
22. Sirunyan AM et al (2020) Performance of the CMS Level-1 trigger in proton-proton collisions at $\sqrt{s}$ = 13 TeV. JINST 15:P10017
23. Tumasyan A et al (2022) Measurement and QCD analysis of double-differential inclusive jet cross sections in pp collisions at $\sqrt{s}$ = 13 TeV. JHEP 02:142

24. Cacciari M, Salam GP, Soyez G (2008) The anti-k_T jet clustering algorithm. JHEP 04:063
25. Cacciari M, Salam GP, Soyez G (2012) Fast jet user manual. Eur Phys J C 72:1896
26. Khachatryan V et al (2017) Jet energy scale and resolution in the CMS experiment in pp collisions at 8 TeV. JINST 12:P02014
27. Jet algorithms performance in 13 TeV data CMS physics analysis summary CMS-PAS-JME-16-003 (2017). https://cdsweb.cern.ch/record/2256875
28. Sirunyan AM et al (2021) Precision luminosity measurement in proton-proton collisions at $\sqrt{s}$ = 13 TeV in 2015 and 2016 at CMS. Eur Phys J C 81:800
29. Khachatryan V et al (2016) Measurement of the double-differential inclusive jet cross section in pro ton-proton collisions at $\sqrt{s}$ = 13 TeV. Eur Phys J C 76:451
30. Chatrchyan S et al (2013) Measurements of differential jet cross sections in proton-proton collisions at $\sqrt{s}$ = 7 TeV with the CMS detector. Phys Rev D 87:112002 [Erratum: Phys Rev D 87:119902 (2013)]
31. Chatrchyan S et al (2014) Measurement of the ratio of inclusive jet cross sections using the Anti-kT algorithm with radius parameters R = 0.5 and 0.7 in pp collisions at $\sqrt{s}$ = 7 TeV. Phys Rev D 90:072006
32. Khachatryan V et al (2017) Measurement and QCD analysis of double-differential inclusive jet cross sections in pp collisions at $\sqrt{s}$ = 8 TeV and cross section ratios to 2.76 and 7 TeV. JHEP 03:156
33. Connor PLS, Žlebčík R (2021) Step: a tool to perform tests of smoothness on differential distributions based on Chebyshev polynomials of the first kind. Submitted to SciPost Phys
34. Agostinelli S et al (2003) G EANT4–a simulation toolkit. Nucl Instrum Meth A 506:250
35. Sjöstrand T et al (2015) An introduction to PYTHIA 8.2. Comput Phys Commun 191:159
36. Khachatryan V et al (2016) Event generator tunes obtained from underlying event and multi-parton scattering measurements. Eur Phys J C 76:155
37. Alwall J, Herquet M, Maltoni F, Mattelaer O, Stelzer T (2011) M ADG RAPH 5: going beyond. JHEP 06:128
38. Alwall J et al (2014) The automated computation of tree-level and next-to-leading order differential cross sections, and their matching to parton shower simulations. JHEP 07:079
39. Sirunyan AM et al (2020) Running of the top quark mass from proton-proton collisions at $\sqrt{s}$ = 13TeV. Phys Lett B 803:135263
40. Sirunyan AM et al (2020) Measurement of tt normalised multi-differential cross sections in pp colli sions at $\sqrt{s}$ = 13 TeV, and simultaneous determination of the strong coupling strength, top quark pole mass, and parton distribution functions. Eur Phys J C 80:658
41. Sirunyan AM et al (2019) Measurement of the tt production cross section, the top quark mass, and the strong coupling constant using dilepton events in pp collisions at $\sqrt{s}$ = 13 TeV. Eur Phys J C 79:368
42. Sirunyan AM et al (2019) Measurements of tt differential cross sections in proton-proton collisions at $\sqrt{s}$ = 13 TeV using events containing two leptons. JHEP 02:149
43. Sirunyan AM et al (2018) Identification of heavy-flavour jets with the CMS detector in pp collisions at 13 TeV. JINST 13:P05011
44. CMS luminosity measurements for the 2016 data taking period technical report. CERN, Geneva (2017). https://cds.cern.ch/record/2257069
45. Cacciari M, Frixione S, Mangano ML, Nason P, Ridolfi G (2004) The t anti-t cross-section at 1.8-TeV and 1.96-TeV: A Study of the systematics due to parton densities and scale dependence. JHEP 04:068
46. Catani S, de Florian D, Grazzini M, Nason P (2003) Soft gluon resummation for Higgs boson production at hadron colliders. JHEP 07:028
47. Skands P, Carrazza S, Rojo J (2014) Tuning PYTHIA 8.1: the Monash 2013 tune. Eur Phys J C 74:3024
48. Investigations of the impact of the parton shower tuning in Pythia 8 in the modelling of tt at $\sqrt{s}$ = 8 and 13 TeV (2016)
49. Dulat S et al (2016) New parton distribution functions from a global analysis of quantum chromodynamics. Phys Rev D 93:033006

50. Bowler MG (1981) e+ e− production of heavy quarks in the string model. Z Phys C 11:169
51. Peterson C, Schlatter D, Schmitt I, Zerwas PM (1983) Scaling violations in inclusive e^+e^- annihilation spectra. Phys Rev D 27:105–111 (1 Jan 1983)
52. Argyropoulos S, Sjöstrand T (2014) Effects of color reconnection on $t\bar{t}$ final states at the LHC. JHEP 11:043
53. Christiansen JR, Skands PZ (2015) String formation beyond leading colour. JHEP 08:003
54. Sirunyan AM et al (2019) Measurement of the top quark Yukawa coupling from $t\bar{t}$ kinematic distributions in the lepton+jets final state in proton-proton collisions at $\sqrt{s} = 13$ TeV. Phys Rev D 100:072007
55. Sirunyan AM et al (2019) Measurement of the top quark polarization and $t\bar{t}$ spin correlations using dilepton final states in proton-proton collisions at $\sqrt{s} = 13$ TeV. Phys Rev D 100:072002
56. Sirunyan AM et al (2018) Measurement of normalized differential tt cross sections in the dilepton channel from pp collisions at $\sqrt{s} = 13$ TeV. JHEP 04:060
57. Khachatryan V et al (2017) Measurement of differential cross sections for top quark pair production using the lepton+jets final state in proton-proton collisions at 13 TeV. Phys. Rev. D 95:092001
58. Sirunyan AM et al (2019) Measurement of the tt production cross section, the top quark mass, and the strong coupling constant using dilepton events in pp collisions at $\sqrt{s} = 13$ TeV. Eur Phys J C 79:368
59. Alioli S et al (2013) A new observable to measure the top-quark mass at hadron colliders. Eur Phys J C 73:2438

Chapter 5
Extraction of m_t^{MSR} Using CMS Data

This chapter presents the implementation of the MSR top quark mass renormalization scheme to the HATHOR [1] inclusive hadronic top quark pair production cross section calculation at NNLO, and the MCFM version 6.8 [2, 3] for single-differential hadronic top quark pair production cross section computations at NLO. Both the MSRn and MSRp schemes are included in the codes. The modified programs take the scale R and the MSR mass $m_t^{MSR}(R)$ as input arguments or, alternatively, an $\overline{MS}$ top quark mass $\overline{m}_t(\overline{m}_t)$, which is subsequently converted into the chosen MSR scheme and evolved to the given R as described in Sect. 2.6. The propagation of the evolved mass to the cross section computation is presented, and the numerical results are demonstrated. Throughout the chapter, 2-loop (3-loop) evolution is implied for NLO (NNLO) calculations. The first study of the behavior of the R scale, and a theoretically consistent extraction of the top quark MSR mass using $t\bar{t}$ pair production cross section data measured as a function of the $t\bar{t}$ invariant mass, are presented.

The inclusive top quark pair production cross section computation implemented up to NNLO in HATHOR is presented in Sect. 5.1. Section 5.2 discusses the implementation and validation of the the single differential top quark pair production cross section at NLO in the MCFM computation, along with studies of the mass renormalization scale independent of the QCD factorization and renormalization scales. Finally, Sect. 5.3 presents the extraction of the top quark MSRn mass by using the single-differential $t\bar{t}$ production cross section measured by the CMS Collaboration in pp collisions at $\sqrt{s} = 13$ TeV [4].

5.1 Inclusive Hadronic Top Quark Pair Production Cross Section at NLO and NNLO

HATHOR is a program for calculating the total cross section for top quark pair production in hadronic collisions up to NNLO. It has also been extended for single top production up to NLO, although this section discusses the computation of the

T. Mäkelä, *Towards Global Interpretation of LHC Data*, Springer Theses,
https://doi.org/10.1007/978-3-031-29779-3_5

hadronic top quark pair production cross section with focus on utilizing the MSRn and MSRp schemes described in Sect. 2.6.

In HATHOR, the inclusive cross section computation relies on the exact results of [5–8], and the electroweak corrections based on [9–11]. Numerical integration is performed by an adopted version of the Vegas integrator [12, 13], and random numbers are generated with Ranlux [14]. The FF library [15] is used for calculating one-loop integrals for single-top production and EW corrections.

The cross section is first computed in the pole mass scheme, and transformed into the running mass schemes. The original HATHOR code includes the option of translating the results into the $\overline{\mathrm{MS}}$ scheme [1]. The present implementation of the MSR schemes follows a similar approach. In the pole mass scheme, the inclusive $\mathrm{t\bar{t}}$ production cross section in a collision of two hadrons is separated into LO, NLO and NNLO contributions as

$$\sigma = (a_S(\mu_r))^2\,\sigma^{(0)}\left(m_t^{\mathrm{pole}}\right) + (a_S(\mu_r))^3\,\sigma^{(1)}\left(m_t^{\mathrm{pole}}\right) + (a_S(\mu_r))^4\,\sigma^{(2)}\left(m_t^{\mathrm{pole}}\right). \quad (5.1)$$

The relation of the pole and MSR masses given in Eq. (2.38) is inserted into the $\sigma^{(i)}$ in Eq. (5.1), which are then expanded in $a_S(R)$, defined in Eq. (2.27). Keeping terms up to $\mathcal{O}(a_S^4)$, the cross section becomes

$$\begin{aligned}
\sigma = {} & (a_S(R))^2\,\sigma^{(0)}\left(m_t^{\mathrm{MSR}}\right) \\
& + (a_S(R))^3\,\sigma^{(1)}\left(m_t^{\mathrm{MSR}}\right) \\
& + (a_S(R))^4\,\sigma^{(2)}\left(m_t^{\mathrm{MSR}}\right) \\
& + (a_S(R))^3\,R d_1^{\mathrm{MSR}} \left.\frac{d\sigma^{(0)}}{dm_t}\right|_{m_t=m_t^{\mathrm{MSR}},\mu_r=R} \\
& + (a_S(R))^4 \left(R d_2^{\mathrm{MSR}} \frac{d\sigma^{(0)}}{dm_t} \right. \\
& \qquad + R d_1^{\mathrm{MSR}} \frac{d\sigma^{(1)}}{dm_t} \\
& \qquad \left.\left. + \frac{R^2}{2} \left(d_1^{\mathrm{MSR}}\right)^2 \frac{d^2\sigma^{(0)}}{dm_t^2} \right)\right|_{m_t=m_t^{\mathrm{MSR}},\mu_r=R} . \qquad (5.2)
\end{aligned}$$

Here each term is initially evaluated at $\mu_r = R$, replacing $a_S(\mu_r)$ by $a_S(R)$ in comparison to Eq. (5.1). The dependence on a general μ_r is restored by setting $\mu = R$ and $\mu' = \mu_r$ in the expansion of Eq. (2.30). Unlike for the decoupling coefficients in the evaluations performed for Eq. (5.2), the logarithms $\log(\mu_r^2/R^2) \equiv L_R$ in Eq. (2.30) do not vanish for general $\mu_r \neq R$. The expansions of each a_S^n are performed retaining only the terms of the appropriate order, ergo at $\mathcal{O}(a_S^4)$ they read

$$(a_S(R))^2 = (a_S(\mu_r))^2 \Big\{ 1 + 2a_S(\mu_r)\beta_0 L_R + (a_S(\mu_r))^2 \left[2\beta_1 L_R + 3\beta_0^2 L_R^2 \right] \Big\}, \tag{5.3}$$

$$(a_S(R))^3 = (a_S(\mu_r))^3 \left[1 + 3a_S(\mu_r)\beta_0 L_R \right], \tag{5.4}$$

$$(a_S(R))^4 = (a_S(\mu_r))^4 . \tag{5.5}$$

Since the MSR scheme is expected to have applications mainly for the top quark mass at scales $0 < R < \overline{m}_t(\overline{m}_t)$, the evolution is replaced by that of the $\overline{\text{MS}}$ scheme with a general mass scale μ_m for $R > \overline{m}_t(\overline{m}_t)$. In this regime, the NNLO cross section is computed as

$$\begin{aligned}
\sigma &= (a_S(\mu_m))^2 \, \sigma^{(0)} \left(\overline{m}_t(\mu_m) \right) \\
&+ (a_S(\mu_m))^3 \, \sigma^{(1)} \left(\overline{m}_t(\mu_m) \right) \\
&+ (a_S(\mu_m))^4 \, \sigma^{(2)} \left(\overline{m}_t(\mu_m) \right) \\
&+ (a_S(\mu_m))^3 \, \overline{m}_t(\mu_m) d_1^{\overline{\text{MS}}}(\mu_m) \left. \frac{d\sigma^{(0)}}{dm_t} \right|_{m_t = \overline{m}_t(\mu_m), \mu_r = \mu_m} \\
&+ (a_S(\mu_m))^4 \left(\overline{m}_t(\mu_m) d_2^{\overline{\text{MS}}}(\mu_m) \frac{d\sigma^{(0)}}{dm_t} \right. \\
&\quad + \overline{m}_t(\mu_m) d_1^{\overline{\text{MS}}}(\mu_m) \frac{d\sigma^{(1)}}{dm_t} \\
&\quad \left. \left. + \frac{(\overline{m}_t(\mu_m))^2}{2} \left(d_1^{\overline{\text{MS}}} \right)^2 \frac{d^2\sigma^{(0)}}{dm_t^2} \right) \right|_{m_t = \overline{m}_t(\mu_m), \mu_r = \mu_m} .
\end{aligned} \tag{5.6}$$

In the regime of $\overline{\text{MS}}$ evolution, the dependence on a general μ_r is again restored by applying Eqs. (5.5)–(5.3), replacing R with μ_m.

The evolution of the MSR mass as a function of R, and the dependence of the inclusive hadronic top quark pair production cross section on R, are depicted in Fig. 5.1. Both the MSRp and MSRn schemes are displayed, and the MSRp mass is observed to be slightly above than the MSRn mass at low values of R. Correspondingly, the cross section is slightly higher for MSRn than MSRp at low R, albeit the difference is small. The results are compared to the NLO and NNLO cross sections computed with the original implementation of the $\overline{\text{MS}}$ scheme in the HATHOR code using the ABMP16 5 flavor PDF [16] and $\overline{m}_t(\overline{m}_t) = 160.68$ GeV. The factorization and renormalization scale uncertainties are obtained for the $\overline{\text{MS}}$ computation by varying μ_r and μ_f by a factor of 1/2 and 2 independently, avoiding the combinations taking $\mu_r/\mu_f \rightarrow 4^{\pm 1}\mu_r/\mu_f$. An envelope of the scale variations is constructed and added to the PDF uncertainty in quadrature, yielding the total uncertainty.

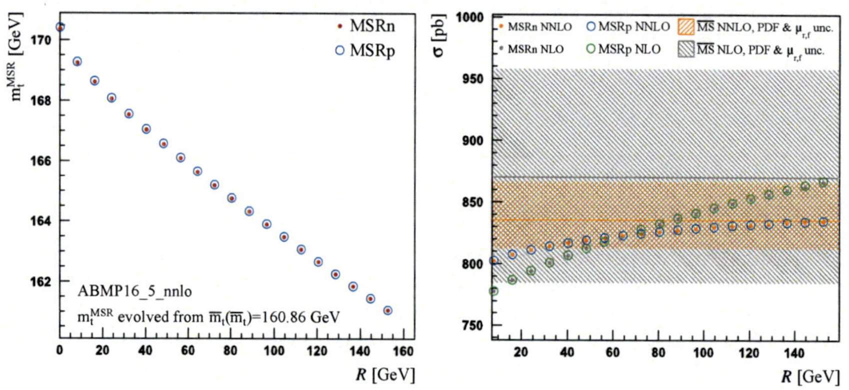

Fig. 5.1 *Left*: the evolution of the MSRp and MSRn masses when R is run from $R = \overline{m}_t = 160.68$ GeV to different values. *Right*: the R-dependence of the inclusive cross section, when the MSR masses are evolved from $\overline{m}_t = 160.68$ GeV to different R values. The shaded bands represent the $\overline{\text{MS}}$ results at NLO and NNLO, with the uncertainties arising from the PDF uncertainty and variations of μ_r and μ_f. The results are produced using the ABMP16 5 flavor PDF [16] at NLO and NNLO, corresponding to the order of the cross section calculation

The implementation of the mass evolution using the MSRn scheme for $R < \overline{m}_t(\overline{m}_t)$ and the $\overline{\text{MS}}$ scheme for $\mu_m > \overline{m}_t(\overline{m}_t)$ is shown at 2 and 3 loops in Fig. 5.2, along with the behavior of independent μ_f and μ_r variations in the NLO and NNLO cross sections in the MSRn scheme, using R values set approximately to $\overline{m}_t(\overline{m}_t)/2$, $\overline{m}_t(\overline{m}_t)$ and $2\overline{m}_t(\overline{m}_t)$, when $\overline{m}_t(\overline{m}_t) = 160.68$ GeV.

A comparison of the conversions between MSR and $\overline{\text{MS}}$ masses shown in Ref. [17], and as calculated at $\mathcal{O}(\alpha_S^3)$ by the presented implementation for HATHOR, is given in Table 5.1. For the present evolution, both the MSRp and MSRn masses are listed.

The results shown in Fig. 5.2 indicate that the difference between the 2 and 3-loop mass evolutions is small, and the deviance becomes noticeable only after evolving the mass scale over intervals close to 100 GeV or more. Furthermore, Fig. 5.2 indicates that the effect of individual μ_r variations in general dominates that of individual μ_f variations, and that the cross sections computed at high values of μ_r and μ_f are subject to larger variation as a function of the MSR ($\overline{\text{MS}}$) scale R (μ_m) than those computed at low values of μ_r and μ_f. Furthermore, the effect of μ_r and μ_f variations is observed to be significantly smaller for the NNLO cross section than for the NLO cross section. However, the computation of the inclusive $t\bar{t}$ production cross section at NLO serves as an important cross-check for the implementation of the single-differential $t\bar{t}$ production cross section, which is discussed in the following.

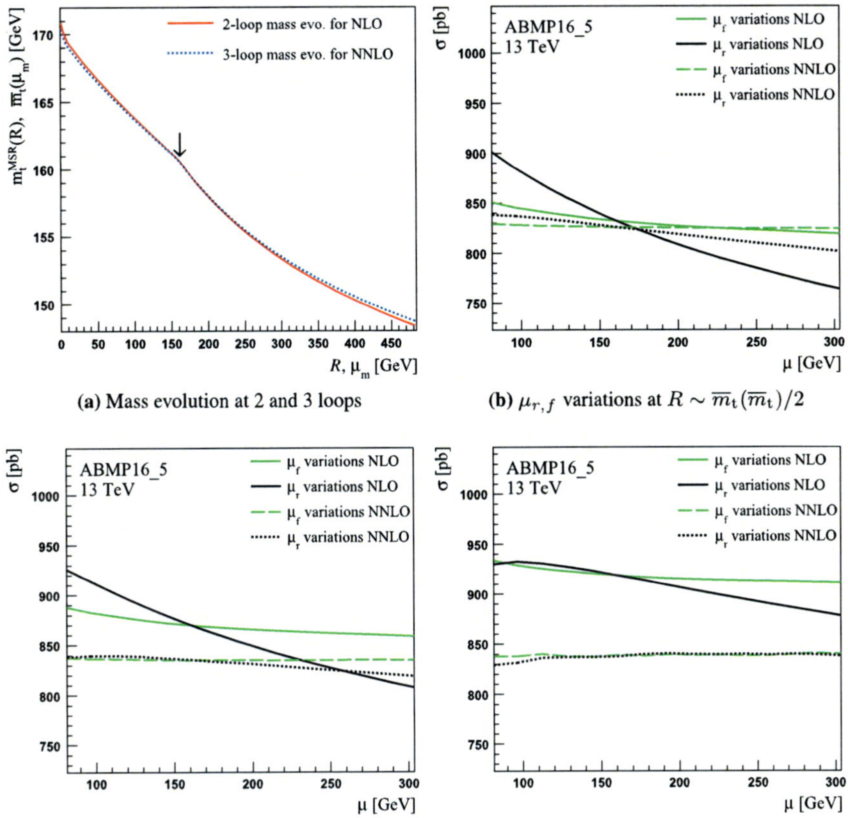

(a) Mass evolution at 2 and 3 loops

(b) $\mu_{r,f}$ variations at $R \sim \overline{m}_t(\overline{m}_t)/2$

(c) $\mu_{r,f}$ variations at $R \sim \overline{m}_t(\overline{m}_t)$

(d) $\mu_{r,f}$ variations at $\mu_m \sim 2\overline{m}_t(\overline{m}_t)$

Fig. 5.2 A comparison of the implemented MSRn and $\overline{\text{MS}}$ mass evolution at 2 and 3 loops, used in the NLO and NNLO cross section computations, respectively, with the transition between the two evolution regimes indicated by an arrow **(a)**. The behavior of independent μ_f and μ_r variations in the NLO and NNLO cross sections in the MSRn scheme with R close to $\overline{m}_t(\overline{m}_t)/2$ **(b)**, $\overline{m}_t(\overline{m}_t)$ **(c)** and $2\overline{m}_t(\overline{m}_t)$ **(d)**. All plots are obtained using the ABMP16 5 flavor PDF at NLO and NNLO, corresponding to the order of the calculation, and by starting the R and μ_m evolution at $\overline{m}_t(\overline{m}_t) = 160.68$ GeV

5.2 Differential Hadronic Top Quark Pair Production Cross Section at NLO

MCFM is a parton-level Monte Carlo program, producing NLO predictions for a multitude of processes at hadron colliders. This section discusses the modification of MCFM version 6.8 [2, 3] to include the implementation of the MSRn and MSRp schemes in the single-differential computation of the hadronic $t\bar{t}$ production cross section. The modifications are based on the procedure presented in Ref. [18].

Table 5.1 For the HATHOR implementation, the MSR(R) columns give the MSR mass obtained by matching the input $\overline{\text{MS}}$ masses in the leftmost to MSR, using the matching relation of Eq. (2.41), and then evolving the mass to R. In contrast, in the columns corresponding to Ref. [17], the MSR masses are evolved from $R = 3$ GeV to the other values. All values are given in units of GeV

	Tab. 2.1 of [17], $m_t^{MSR}(3) \to$ others			Reproduced, $\overline{\text{MS}} \to$MSRp				Reproduced, $\overline{\text{MS}} \to$MSRn		
$\overline{\text{MS}}$	MSR(1)	MSR(3)	MSR(9)	MSR(1)	MSR(3)	MSR(9)	MSR($\overline{m}_t$)	MSR(1)	MSR(3)	MSR(9)
162.62	172.52	172.20	171.58	172.513	172.196	171.579	162.654	172.503	172.190	171.577
162.81	172.72	172.40	171.78	172.713	172.396	171.778	162.844	172.703	172.389	171.777
163.00	172.92	172.60	171.98	172.912	172.595	171.978	163.034	172.902	172.589	171.976
163.19	173.12	172.80	172.18	173.112	172.795	172.177	163.224	173.102	172.788	172.176
163.38	173.32	173.00	172.38	173.311	172.994	172.377	163.414	173.301	172.988	172.375
163.57	173.52	173.20	172.58	173.511	173.194	172.577	163.604	173.501	173.188	172.575
163.76	173.72	173.40	172.78	173.710	173.394	172.776	163.794	173.700	173.387	172.774
163.95	173.92	173.60	172.98	173.910	173.593	172.976	163.984	173.900	173.587	172.974
164.14	174.12	173.80	173.18	174.109	173.793	173.175	164.174	174.099	173.786	173.173
164.33	174.32	174.00	173.38	174.309	173.992	173.375	164.364	174.299	173.986	173.373
164.52	174.52	174.20	173.58	174.509	174.192	173.574	164.554	174.498	174.185	173.572

In the pole mass scheme, the single-differential cross section is separated into the LO and NLO contributions as

$$\frac{d\sigma}{dX} = (a_S(\mu_r))^2 \frac{d\sigma^{(0)}}{dX}\left(m_t^{\text{pole}}, \mu_r, \mu_f\right) + (a_S(\mu_r))^3 \frac{d\sigma^{(1)}}{dX}\left(m_t^{\text{pole}}, \mu_r, \mu_f\right). \quad (5.7)$$

In contrast to the NNLO cross section procedure discussed in Sect. 5.1, it is straightforward to keep explicit μ_r dependence throughout the computation when working at $\mathcal{O}(a_S^3)$, and the differential NLO cross section in the MSR scheme becomes

$$\begin{aligned}\frac{d\sigma}{dX} &= (a_S(\mu_r))^2 \frac{d\sigma^{(0)}}{dX}\left(m_t^{\text{MSR}}, \mu_r, \mu_f\right) + (a_S(\mu_r))^3 \frac{d\sigma^{(1)}}{dX}\left(m_t^{\text{MSR}}, \mu_r, \mu_f\right) \\ &+ (a_S(\mu_r))^3\, R d_1^{\text{MSR}} \frac{d}{dm_t}\left(\frac{d\sigma^{(0)}(m_t, \mu_r, \mu_f)}{dX}\right)\Bigg|_{m_t = m_t^{\text{MSR}}}. \end{aligned} \quad (5.8)$$

Similarly, in the region $\mu_m > \overline{m}_t(\overline{m}_t)$, the cross section reads

$$\begin{aligned}\frac{d\sigma}{dX} &= (a_S(\mu_r))^2 \frac{d\sigma^{(0)}}{dX}\left(\overline{m}_t(\mu_m), \mu_r, \mu_f\right) + (a_S(\mu_r))^3 \frac{d\sigma^{(1)}}{dX}\left(\overline{m}_t(\mu_m), \mu_r, \mu_f\right) \\ &+ (a_S(\mu_r))^3\, \overline{m}_t(\mu_m) d_1^{\overline{\text{MS}}}(\mu_m) \frac{d}{dm_t}\left(\frac{d\sigma^{(1)}(m_t, \mu_r, \mu_f)}{dX}\right)\Bigg|_{m_t = \overline{m}_t(\mu_m)}. \end{aligned} \quad (5.9)$$

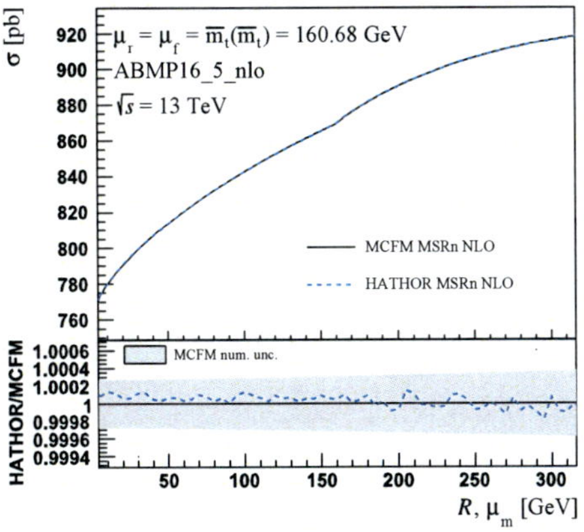

Fig. 5.3 A comparison of the inclusive hadronic top quark pair production cross sections obtained using the MSRn scheme as implemented in the MCFM (black) and HATHOR (blue) computations. The ratio plot shows the HATHOR cross section divided by the MCFM cross section (blue) and the fractional MCFM numerical uncertainty. The plots are obtained using NLO evolution, the ABMP16 5 flavor PDF at NLO and setting μ_r, μ_f equal to the $\overline{\text{MS}}$ top quark mass $\overline{m}_\text{t}(\overline{m}_\text{t}) = 160.68$ GeV

5.2.1 Validation of the MCFM Implementation

A comparison of the inclusive cross sections computed using the HATHOR and MCFM implementations of the MSRn scheme is shown in Fig. 5.3. All differences are observed to be within the numerical uncertainties of the MCFM result.

The single-differential NLO cross section results obtained with the MCFM implementation are further compared to an external code translating the standard MCFM pole mass scheme cross section results into the MSRn scheme. Contributions to the cross sections are broken down to the LO term, virtual and real emission corrections to the NLO term, and the derivative term which is on the second line in Eqs. (5.8) and (5.9). The comparison is shown in Fig. 5.4 for the $m_{\text{t}\bar{\text{t}}}$ distribution, in Fig. 5.5 for the $p_\text{T}(\bar{\text{t}})$ distribution, and in Fig. 5.6 for the $y(\bar{\text{t}})$ distribution. Very good agreement between the two approaches is observed in all cases.

The implementation of the single-differential cross section allows investigating the hitherto unexamined behavior of the scale R in a way that distinguishes it from the scales μ_r and μ_f. This is particularly important for enhancing the precision in extracting the top quark mass from the $m_{\text{t}\bar{\text{t}}}$ distribution by shedding new light on the estimation of scale uncertainties.

5.2.2 R Behavior in the Differential Case

To examine the dependence of the $m_{\text{t}\bar{\text{t}}}$ distribution on the scales μ_r, μ_f and R or μ_m, depending on whether MSRn or $\overline{\text{MS}}$ evolution is used, the cross section predictions

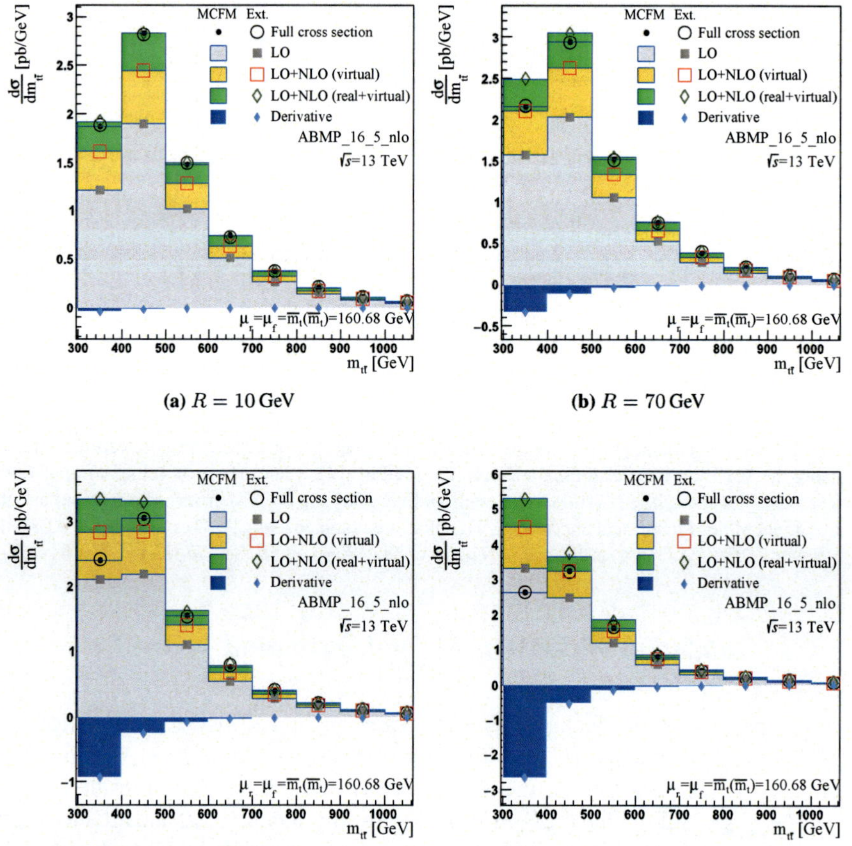

(a) $R = 10\,\text{GeV}$ (b) $R = 70\,\text{GeV}$

(c) $R = 160\,\text{GeV}$ (d) $\mu_m = 320\,\text{GeV}$

Fig. 5.4 The contributions to the single-differential cross section in bins of the $t\bar{t}$ invariant mass. The black dots are the total result of the MSRn cross section and the filled histograms the various contributions to it, as obtained from the MCFM implementation. The open circles indicate the total result and the colored markers the different contributions, as obtained from the external validation code

are investigated using a finer binning in $m_{t\bar{t}}$, corresponding to bin limits from 300 GeV to about 1200 GeV, when the width of each bin is 33 GeV. The binning is illustrated by the breakdown of the different contributions to the NLO cross section in Fig. 5.7.

Figure 5.8 shows the LO and NLO cross sections for the six bins in the range $300\,\text{GeV} < m_{t\bar{t}} < 498\,\text{GeV}$ as a function of R (μ_m) in the MSR ($\overline{\text{MS}}$) mass evolution regime. The effect of the scales μ_r and μ_f is investigated by setting them to $\overline{m}_t(\overline{m}_t)$ multiplied by factors of 1/4, 1/2, 1, 2 and 4. The six bins included in the range $498\,\text{GeV} < m_{t\bar{t}} < 696\,\text{GeV}$ are shown in Fig. 5.9. At higher values of $m_{t\bar{t}}$, the cross section becomes small and shows no significant dependence on the scales R and μ_m,

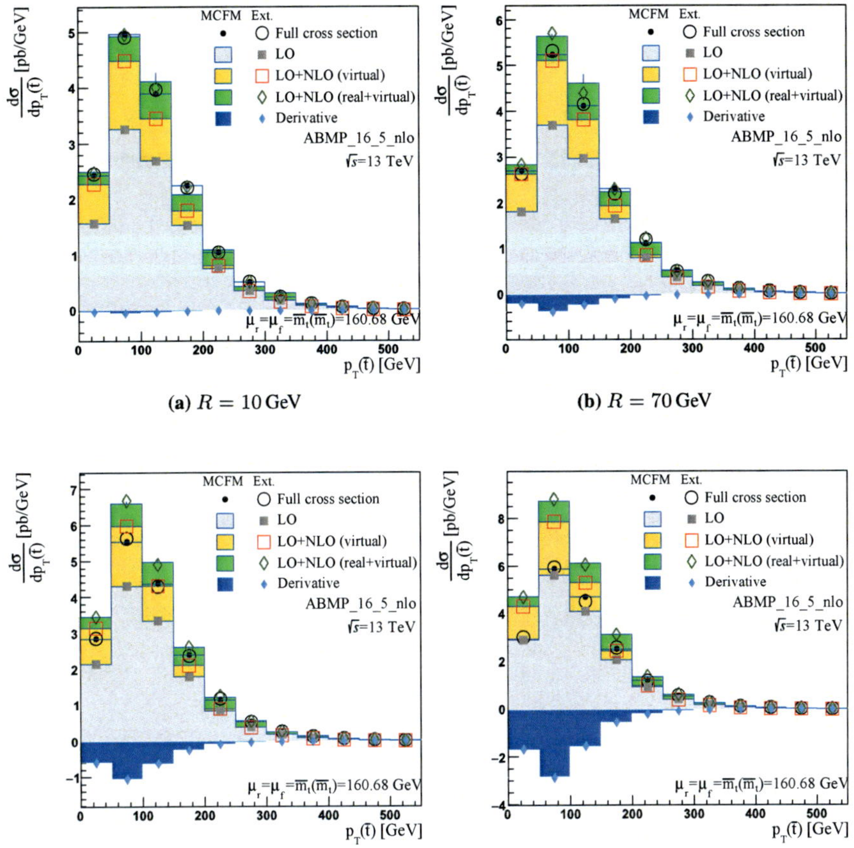

(a) $R = 10\,\text{GeV}$ (b) $R = 70\,\text{GeV}$

(c) $R = 160\,\text{GeV}$ (d) $\mu_m = 320\,\text{GeV}$

Fig. 5.5 The contributions to the single-differential cross section in bins of the transverse momentum of the top antiquark. The black dots are the total result of the MSRn cross section and the filled histograms the various contributions to it, as obtained from the MCFM implementation. The open circles indicate the total result and the colored markers the different contributions, as obtained from the external validation code

regardless of any factors applied on μ_r and μ_f. The LO and NLO cross sections are presented in Figs. 5.8 and 5.9 and the ratios of the NLO and LO cross sections are shown in Figs. 5.10 and 5.11.

The cross section vanishes in the $m_{t\bar{t}} \in [300, 333]\,\text{GeV}$ bin when $R \lesssim 60\,\text{GeV}$, which corresponds to $2m_t^{\text{MSR}}(R) > 333\,\text{GeV}$ as highlighted in Fig. 5.12. Non-zero contributions to the cross section in the $m_{t\bar{t}} \in [300, 333]\,\text{GeV}$ bin appear only at sufficiently high values of R or μ_m, corresponding to smaller values of $m_t^{\text{MSR}}(R)$ or $\overline{m}_t(\mu_m)$. As expected, the LO contribution to the cross section is zero or positive for all values of R and μ_m. However, the NLO cross section becomes negative, since

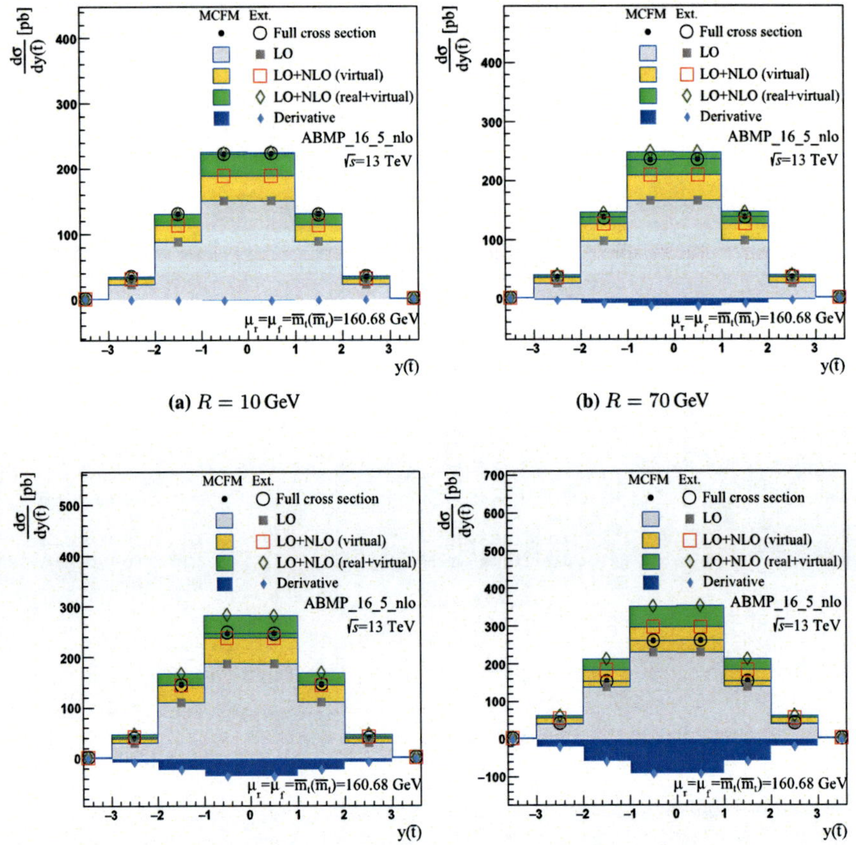

(a) $R = 10\,\text{GeV}$ (b) $R = 70\,\text{GeV}$

(c) $R = 160\,\text{GeV}$ (d) $\mu_m = 320\,\text{GeV}$

Fig. 5.6 The contributions to the single-differential cross section in bins of the rapidity of the top antiquark. The black dots are the total result of the MSRn cross section and the filled histograms the various contributions to it, as obtained from the MCFM implementation. The open circles indicate the total result and the colored markers the different contributions, as obtained from the external validation code

the derivative terms in Eqs. (5.8)–(5.9) decrease quicker than the other contributions increase. This feature can be attributed to fixed-order pQCD lacking a proper treatment for the *Coulomb singularity* near the $t\bar{t}$ production threshold. At energies close to the $t\bar{t}$ production threshold, the produced top quarks attain small non-relativistic velocities $v \ll 1$. The dynamics of the $t\bar{t}$ system are hence governed by the mass m_t, the relative momentum $m_t v$ and the kinetic energy $m_t v^2$ of the top quark. Since $m_t \gg m_t v \gg m_t v^2$, the appearance of ratios involving the masses, momenta and kinetic energy of the top quark renders the standard multi-loop expansion in α_S, on which the cross section calculations rely on, unreliable at the threshold of $t\bar{t}$ production. The most pronounced issue arises from the Coulomb interaction, i.e. the

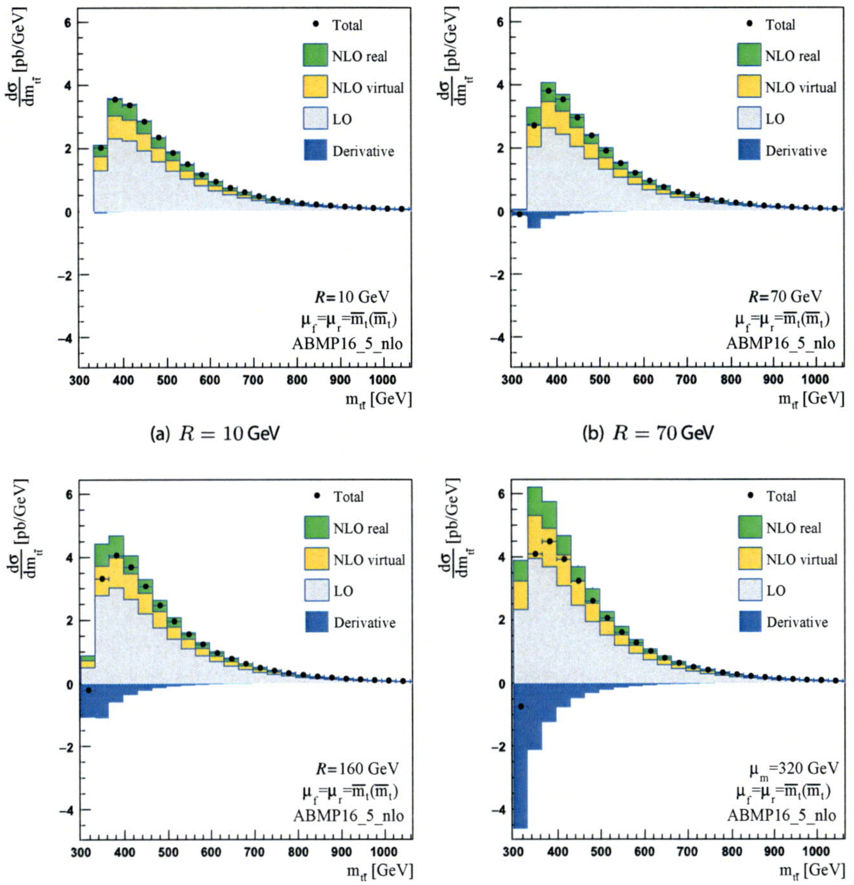

(a) $R = 10$ GeV (b) $R = 70$ GeV (c) $R = 160$ GeV (d) $\mu_m = 320$ GeV

Fig. 5.7 The contributions to the single-differential cross section in bins of the $\mathrm{t\bar{t}}$ invariant mass, using a bin width of 33 GeV. The black dots are the total result of the cross section using the MSRn scheme, and the filled histograms represent the various contributions to it

exchange of a boson between the produced t and $\bar{\mathrm{t}}$, with a prediction depending on the ratio $m_\mathrm{t}/(m_\mathrm{t} v)$. This leads to the appearance of singular $(\alpha_S/v)^n$ behaviour in the n-loop fixed-order pQCD correction [19]. This is to be solved by quasi-boundstate corrections and the resummation of soft gluon effects, following e.g. Ref. [20].

Another noticeable feature of the $m_{\mathrm{t\bar{t}}} \in [300, 333]$ GeV bin shown in Fig. 5.12 is the sudden increase of the cross section at $\mu_m \gtrsim 410$ GeV. This occurs when $m_\mathrm{t}^\mathrm{MSR}$ is small enough for the $\mathrm{t\bar{t}}$ production threshold to be below 300 GeV, and the missing corrections at the threshold are expected to be of less importance. A similar effect is seen in the $m_{\mathrm{t\bar{t}}} \in [333, 366]$ GeV bin, as the threshold effects contribute more to

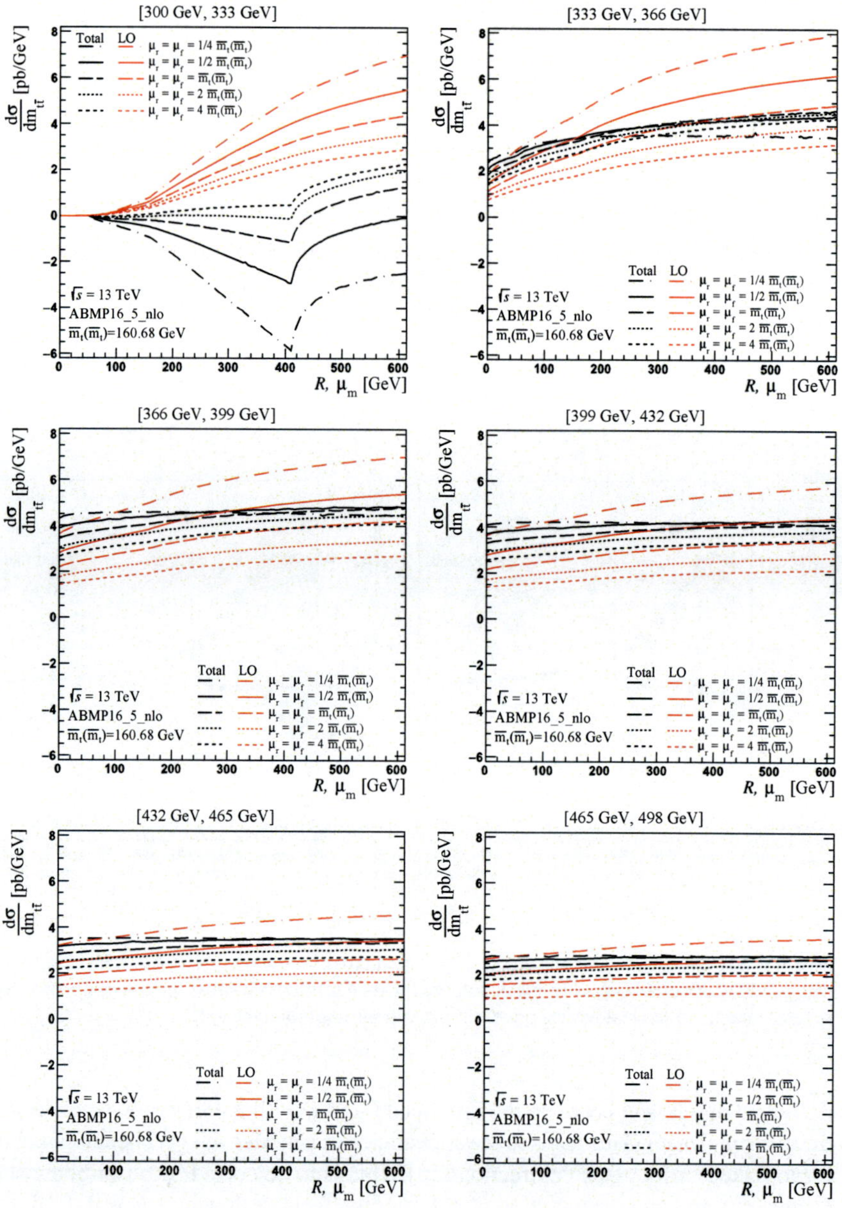

Fig. 5.8 The mass renormalization scale dependence of the first six bins of the $m_{t\bar{t}}$ distribution using the binning introduced Fig. 5.7

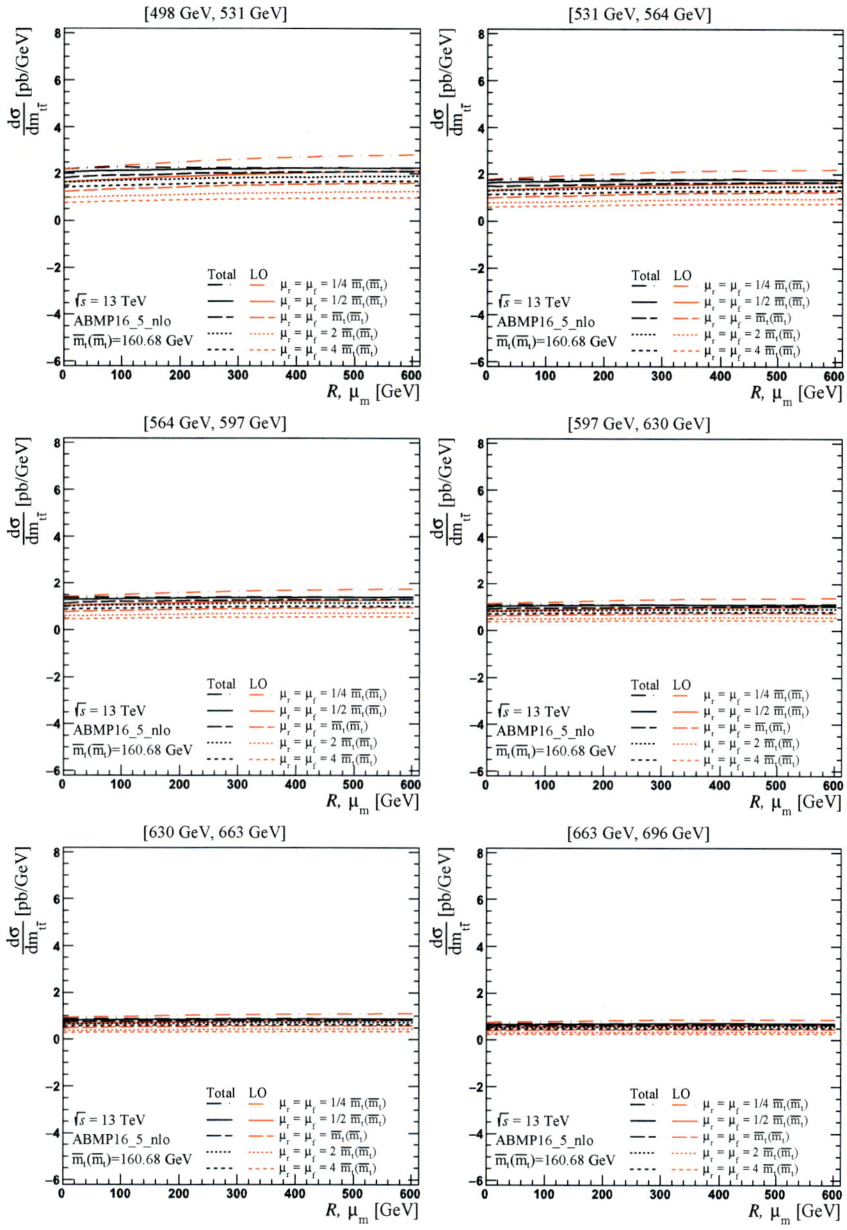

Fig. 5.9 The mass renormalization scale dependence of the six bins following those presented in Fig. 5.8

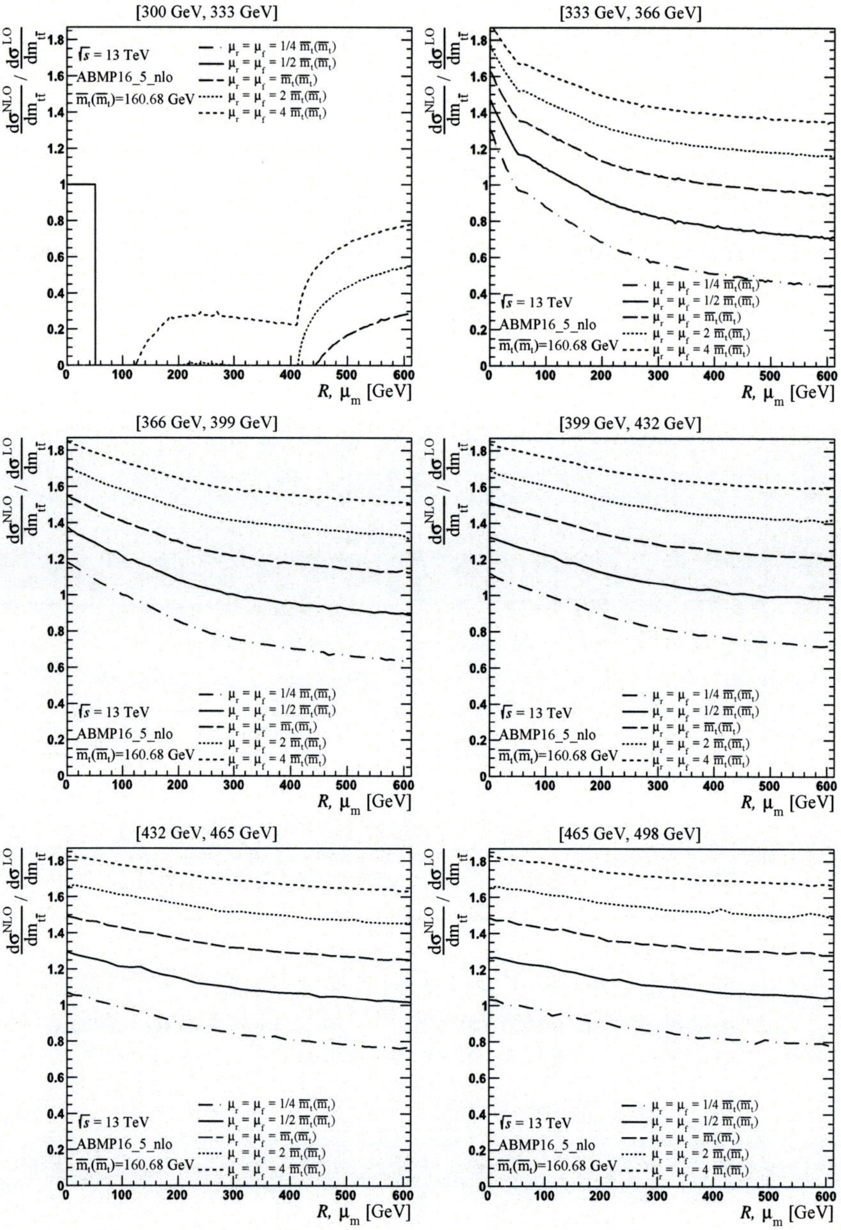

Fig. 5.10 The ratio of the NLO and LO cross sections presented in Fig. 5.8

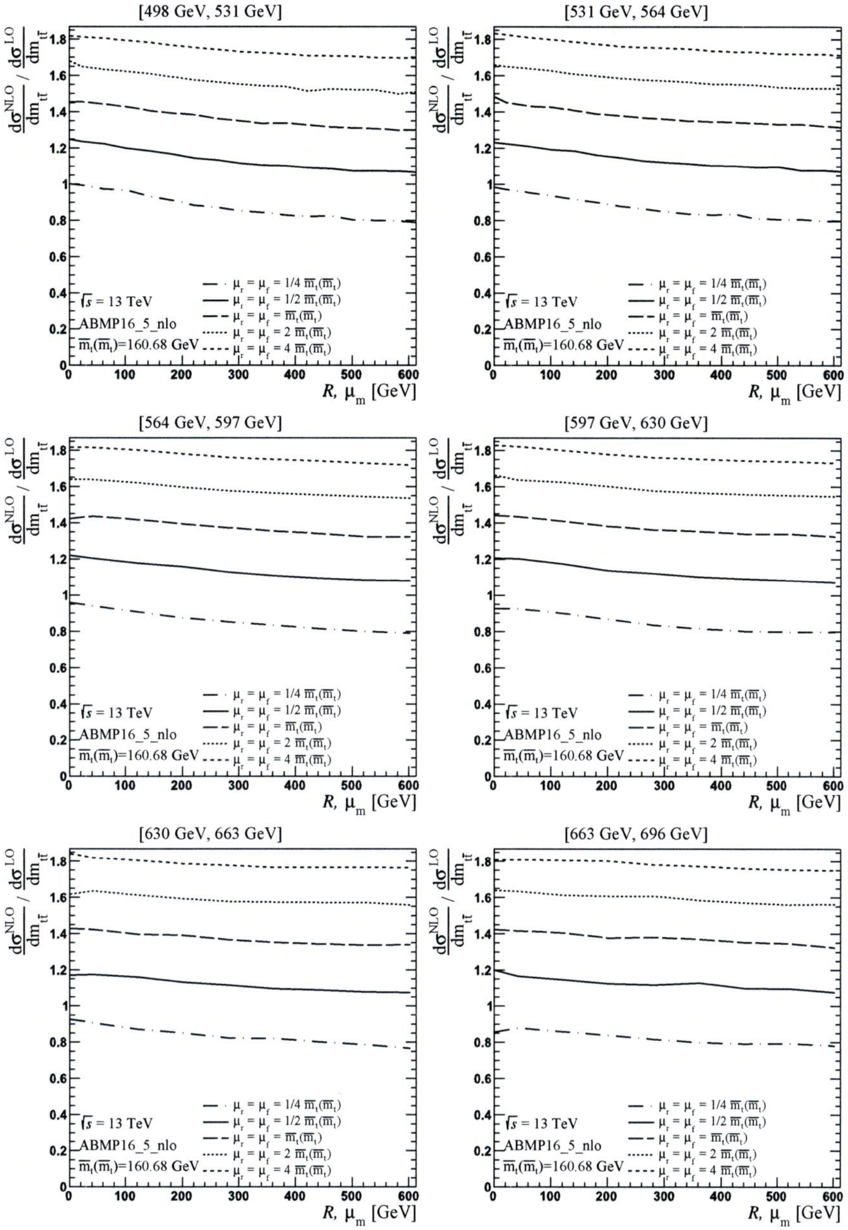

Fig. 5.11 The ratio of the NLO and LO cross sections presented in Fig. 5.9

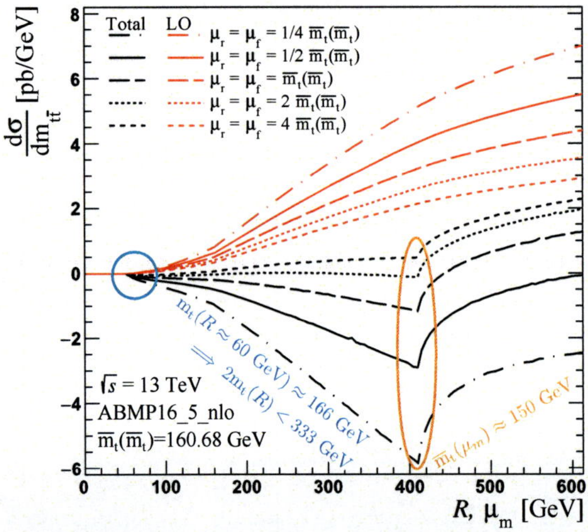

Fig. 5.12 The $m_{t\bar{t}} \in [300, 333]$ GeV bin of the $m_{t\bar{t}}$ distribution introduced in Fig. 5.7. There is no $t\bar{t}$ production at $R \lesssim 60$ GeV, but the region above it suffers from the lack of Coulomb corrections in the MCFM v6.8 code, until $\mu_m \gtrsim 410$ GeV where the threshold for $t\bar{t}$ production is below the lower bin boundary

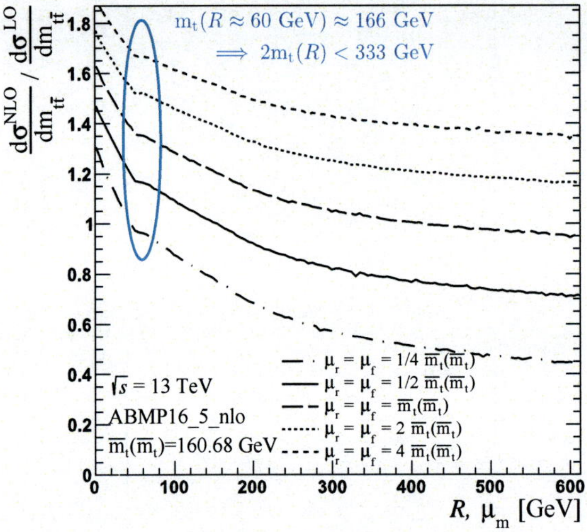

Fig. 5.13 The ratio of the LO and NLO results for the $m_{t\bar{t}} \in [333, 366]$ GeV bin of the $m_{t\bar{t}}$ distribution introduced in Fig. 5.7. The transition from a region suffering from the missing Coulomb corrections to a more stable region where the threshold effects become of less importance is seen at $R \gtrsim 60$ GeV

the $d\sigma/dm_{t\bar{t}}$ cross section in the $R \lesssim 60$ GeV region than at higher R and μ_m. The transition is illustrated in the ratio of the NLO and LO cross sections and highlighted in Fig. 5.13.

Although the results of the $m_{t\bar{t}} \in [300, 333]$ GeV bin are unphysical, the observed features suggest an explanation for the behaviour of the cross section in the $m_{t\bar{t}} \in [333, 366]$ GeV bin, which is close to the peak of the $m_{t\bar{t}}$ distribution and hence sensitive to the top quark mass. Therefore, understanding the scale behavior in this region is particularly important for extractions of $m_t^{\text{MSR}}(R)$ from cross sections measured as a function of $m_{t\bar{t}}$. In general, the NLO cross sections in each bin, excluding $m_{t\bar{t}} \in [300, 333]$ GeV, are observed to stabilize quickly with increasing R

or μ_m when μ_r and μ_f are set to small values, provided that R is high enough for the stabilization of the predicted cross section to start in the $m_{\mathrm{t\bar{t}}} \in [333, 366]$ GeV bin. Therefore, the results suggest setting the central value of the scales μ_r, μ_f and R to a value around $\overline{m}_{\mathrm{t}}(\overline{m}_{\mathrm{t}})/2 \approx 80$ GeV for obtaining predictions that are robust against scale variations.

5.3 Extraction of the Top Quark MSRn Mass at the LHC

As a phenomenological application of the implementation of the MSR schemes in the MCFM v6.8 program, the MSRn mass $m_{\mathrm{t}}^{\mathrm{MSR}}(R)$ is extracted from a measurement of the single-differential $d\sigma/dm_{\mathrm{t\bar{t}}}$ cross section in pp collisions at $\sqrt{s} = 13$ TeV [4], described in Sect. 4.4.

The MCFM predictions for the single-differential $d\sigma/dm_{\mathrm{t\bar{t}}}$ cross section are computed using the ABMP16 5 flavor PDF set [16] at NLO. Initially, the scale R is set to 80 GeV, and the cross section is computed for a range of assumed values of $m_{\mathrm{t}}^{\mathrm{MSR}}(80\,\mathrm{GeV})$. The predictions are then compared to the experimental data, and the function

$$\chi^2 = \sum_{i,j} (\sigma_i^{\mathrm{exp}} - \sigma_i^{\mathrm{th}}) C_{ij}^{-1} (\sigma_j^{\mathrm{exp}} - \sigma_j^{\mathrm{th}}), \tag{5.10}$$

is computed for each $m_{\mathrm{t}}^{\mathrm{MSR}}(80\,\mathrm{GeV})$. Here σ_i^{exp} are the experimental data and σ_i^{th} the theoretical predictions, the indices i, j run over the bins of the distribution, and C_{ij}^{-1} is the inverse covariance matrix. One step of the scan is illustrated by the comparison of the experimental data to the theoretical prediction at $\mu_r = \mu_f = m_{\mathrm{t}}^{\mathrm{MSR}}(80\,\mathrm{GeV}) = 167$ GeV in Fig. 5.14.

Figure 5.15 shows a 4th order polynomial fit to the χ^2 values resulting from the scan, when μ_r and μ_f are set to $\mu = m_{\mathrm{t}}^{\mathrm{MSR}}(80\,\mathrm{GeV})$ for all 4 bins of the $m_{\mathrm{t\bar{t}}}$ distribution. Since the $\mathrm{t\bar{t}}$ production threshold region is subject to the missing Coulomb corrections, an alternative fit is performed excluding the $m_{\mathrm{t\bar{t}}} < 420$ GeV bin, and the resulting χ^2 distribution is shown in Fig. 5.16. The exclusion of the $m_{\mathrm{t\bar{t}}} < 420$ GeV bin leads to a loss of sensitivity and, consequently, large uncertainties in the resulting top quark mass. As discussed in Sect. 5.2.2, lowering the values of μ_r and μ_f stabilizes the predictions as a function of R. Therefore, a third fit is performed by setting μ_r and μ_f to $m_{\mathrm{t}}^{\mathrm{MSR}}(80\,\mathrm{GeV})/2$ for the $m_{\mathrm{t\bar{t}}} < 420$ GeV bin, and to $m_{\mathrm{t}}^{\mathrm{MSR}}(80\,\mathrm{GeV})$ for the remainder. The respective χ^2 scan from this configuration is shown in Fig. 5.17.

The fit uncertainties are obtained by using the $\Delta\chi^2 = 1$ tolerance criterion, while the μ_r and μ_f scale uncertainties are evaluated by varying the central value $\mu^{(i)}$ in each bin $i \in \{1, 2, 3, 4\}$ up and down by a factor of 2, avoiding the cases taking $\mu_r^{(i)}/\mu_f^{(i)} \to 4^{\pm 1} \mu_r^{(i)}/\mu_f^{(i)}$, and constructing an envelope. The extracted mass values are evolved from $R = 80$ GeV to the reference scales of $R \in \{1, 2, 3\}$ GeV.

Note, that in Ref. [22] the reference value of R is chosen as $R = 1$ GeV. This choice however requires evaluating $\alpha_S(1\,\mathrm{GeV})$, which is close to the Landau pole. Using a reference scale of $R = 2$ GeV or higher is expected to become increasingly

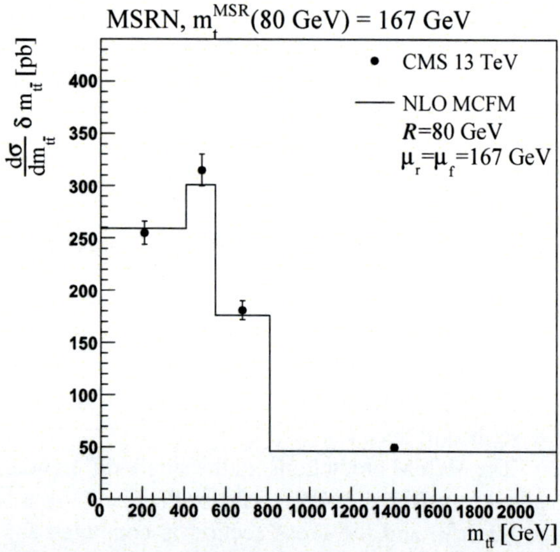

Fig. 5.14 The cross section of $t\bar{t}$ production as a function of $m_{t\bar{t}}$ multiplied by the bin width $\delta m_{t\bar{t}}$, measured in pp collisions at $\sqrt{s} = 13$ TeV by the CMS Collaboration [4] (filled circles). The NLO prediction (histogram) is obtained using MCFM in the MSRn scheme. The scales in the predictions are set to $R = 80$ GeV and $\mu_r = \mu_f = m_t^{MSR}(R) = 167$ GeV for all bins of the $m_{t\bar{t}}$ distribution

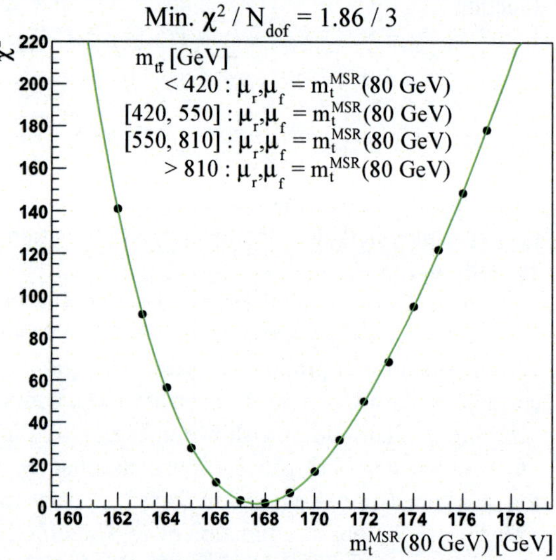

Fig. 5.15 A 4th order polynomial fitted to the χ^2 resulting from comparing the experimental data to theory predictions assuming different values of $m_t^{MSR}(80\,\text{GeV})$. The scales μ_r and μ_f are set to $m_t^{MSR}(80\,\text{GeV})$ for all four bins of the $m_{t\bar{t}}$ distribution [21]

important in future extractions of the MSR mass, since $\alpha_S(\mu)$ may become unstable as $\mu \to 1$ GeV when using higher-loop evolution for α_S. To ensure the stability of the result, the mass extracted here is also reported at $R = 2$ GeV and $R = 3$ GeV. Furthermore, the $m_t^{MSR}(80\,\text{GeV})$ value is translated into the $\overline{\text{MS}}$ mass by finding $m_t^{MSR}(m_t^{MSR})$ value iteratively via the condition $R = m_t^{MSR}(R)$, and applying the matching formula given in Eq. (2.42), up to $\mathcal{O}(a_S^3)$. The uncertainty of the initial choice of R is assessed by repeating the fits at $R = 60$ GeV and $R = 100$ GeV, and

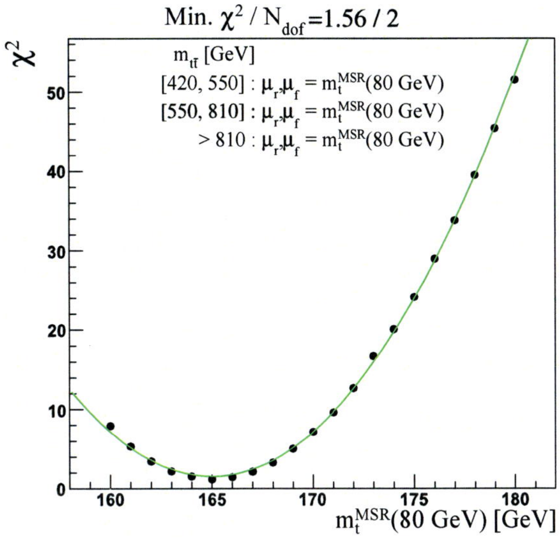

Fig. 5.16 A 4th order polynomial fitted to the χ^2 resulting from comparing the experimental data to theory predictions assuming different values of $m_t^{MSR}(80\,\text{GeV})$, when the $m_{t\bar{t}} < 420\,\text{GeV}$ bin is excluded. The scales μ_r and μ_f are set to $m_t^{MSR}(80\,\text{GeV})$ for the remaining bins

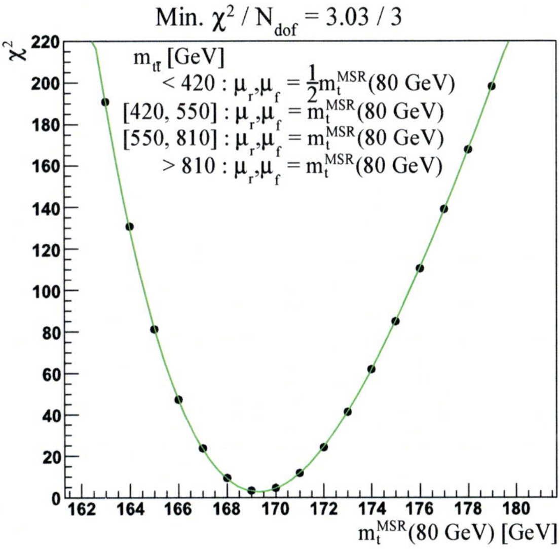

Fig. 5.17 A 4th order polynomial fitted to the χ^2 resulting from comparing the experimental data to theory predictions assuming different values of $m_t^{MSR}(80\,\text{GeV})$. The scales μ_r and μ_f are set to $m_t^{MSR}(80\,\text{GeV})/2$ for the $m_{t\bar{t}} < 420\,\text{GeV}$ bin, and to $m_t^{MSR}(80\,\text{GeV})$ for the remainder [21]

taking the difference in the resulting $m_t^{MSR}(1\,\text{GeV})$, $m_t^{MSR}(2\,\text{GeV})$ and $m_t^{MSR}(3\,\text{GeV})$ to the respective values obtained in the $R = 80\,\text{GeV}$ fit as the R scale uncertainty.

The resulting values for the top quark MSRn mass are listed in Table 5.2. In particular, setting the central μ_r and μ_f to $m_t^{MSR}(80\,\text{GeV})$ considering the complete $m_{t\bar{t}}$ distribution results in an extracted value of $m_t^{MSR}(1\,\text{GeV})$ compatible with the results[1] of Ref. [22]. Furthermore, translating the $m_t^{MSR}(80\,\text{GeV})$ obtained in this fit

[1] For reference, the results of Ref. [22] are reported in Eqs. (2.51)–(2.52) of this thesis.

Table 5.2 The values of $m_t^{MSR}(R)$ obtained at different scales R (given in brackets below m_t^{MSR}), and the corresponding $\overline{m}_t(\overline{m}_t)$, along with the fit and scale uncertainties for the $m_t^{MSR}(R)$ extracted at $R = 80\,\text{GeV}$. The leftmost column gives the central choice of μ_r and μ_f for each bin in the distribution, when $\mu = m_t^{MSR}(80\,\text{GeV})$, and unused bins are indicated by a dash. The fit and μ_r, μ_f uncertainties correspond to the MSRn mass extracted at $R = 80\,\text{GeV}$. Performing the evolution from $m_t^{MSR}(80\,\text{GeV})$ to $m_t^{MSR}(1\,\text{GeV})$, $m_t^{MSR}(2\,\text{GeV})$, $m_t^{MSR}(3\,\text{GeV})$ or $\overline{m}_t(\overline{m}_t)$ results in an R uncertainty that agrees in all cases within the given accuracy, and is reported in the rightmost column

	m_t^{MSR} (80 GeV) [GeV]	m_t^{MSR} (1 GeV) [GeV]	m_t^{MSR} (2 GeV) [GeV]	m_t^{MSR} (3 GeV) [GeV]	$\overline{m}_t(\overline{m}_t)$ [GeV]	Fit [GeV]	(μ_r, μ_f) [GeV]	(R) [GeV]
(μ, μ, μ, μ)	167.7	173.2	173.0	172.9	163.3	$^{+0.6}_{-0.6}$	$^{+0.4}_{-0.6}$	$^{+0.4}_{-0.5}$
$(-, \mu, \mu, \mu)$	165.0	170.5	170.3	170.2	160.7	$^{+2.1}_{-2.1}$	$^{+6.7}_{-9.8}$	$^{+0.5}_{-0.5}$
$(\frac{\mu}{2}, \mu, \mu, \mu)$	169.3	174.8	174.6	174.5	164.8	$^{+0.5}_{-0.5}$	$^{+0.2}_{-0.4}$	$^{+0.2}_{-0.3}$

into $\overline{m}_t(\overline{m}_t)$ yields a result agreeing with Ref. [4]. Removing the $m_{t\bar{t}} < 420\,\text{GeV}$ bin is observed to reduce the value of the extracted top quark mass, but it also increases the uncertainties by an order of magnitude.

In accordance with the results shown in Fig. 5.8, multiplying the scales μ_r and μ_f by 1/2 in the $m_{t\bar{t}} < 420\,\text{GeV}$ bin results in an increase in the value of the NLO cross section when $R = 80\,\text{GeV}$. The fit for $m_t^{MSR}(80\,\text{GeV})$ compensates for this effect by preferring larger values of the top quark MSR mass, which brings the predicted cross section down, especially near the threshold region. As expected from the observations in Sect. 5.2.2, this scale setting yields smaller scale uncertainties than the other configurations investigated in this thesis. The effect is attributed to increased robustness against scale variations. Evolving the resulting value of

$$m_t^{MSRn}(80\,\text{GeV}) = 169.3 \pm 0.5\,(\text{fit})^{+0.2}_{-0.4}\,(\mu_r, \mu_f)^{+0.2}_{-0.3}\,(R)\,\text{GeV} \tag{5.11}$$

to the reference scale $R = 3\,\text{GeV}$ and adding all the uncertainties in quadrature to obtain a total uncertainty, the obtained value of $m_t^{MSR}(3\,\text{GeV}) = 174.5^{+0.6}_{-0.7}\,\text{GeV}$ is observed to be in tension with the $m_t^{MSR}(3\,\text{GeV}) = 169.6^{+0.8}_{-1.1}\,\text{GeV}$ extracted in Ref. [23]. As notable dissimilarities, the selection of experimental data in the present work is different to Ref. [23], where $m_t^{MSR}(3\,\text{GeV})$ is extracted simultaneously with the PDFs and strong coupling constant, resulting in the value $\alpha_S(m_Z) = 0.1132^{+0.0023}_{-0.0018}$, which is within 2 standard deviations from the value of the ABMP16 fit at NLO [23]. Furthermore, the computation of the cross section in the MSR scheme is not implemented directly into the MC calculation in Ref. [23], but the theoretical predictions are obtained in the pole mass scheme using MADGRAPH5_aMC@NLO [24, 25] interfaced with AMCFAST [26] and APPLGRID [27], and then translated to the MSR scheme at $R = 3\,\text{GeV}$ via Eq. (2.54). Notably, neither Ref. [23] nor the present work include a correction for the Coulomb effects, but the extraction of the top quark MSR mass presented in this thesis is performed using predictions in the MSR scheme at $R = 80\,\text{GeV}$ to optimize the stability of the cross section. The resulting $m_t^{\text{MSR}}(80\,\text{GeV})$ is then evolved to the reference R scales, in contrast to performing the extraction at $R = 3\,\text{GeV}$. Moreover, the computation in Ref. [23] relies on the MSRp scheme instead of the MSRn scheme used in this thesis, although the contribution from the choice of scheme is expected to be small at $R = 3\,\text{GeV}$ as indicated by Table 5.1.

In this thesis, the first examination of the behavior of the R scale is presented, along with an extraction of the top quark MSR mass. The results obtained by setting $\mu_r = \mu_f = m_t^{\text{MSR}}(80\,\text{GeV})$, considering the complete $m_{t\bar{t}}$ distribution, agree with the previous studies of the running of the top quark $\overline{\text{MS}}$ mass [4] and the top quark MSR mass performed by comparing NLL predictions to simulated MC samples [22]. Applying low values of μ_r and μ_f, particularly in the vicinity of the $t\bar{t}$ production threshold region in the $m_{t\bar{t}}$ distribution, is observed to decrease the scale uncertainty in the determination of the MSR mass. Therefore, the use of dynamical scales yielding low values in the low-$m_{t\bar{t}}$ regime is recommended for future extractions of the top quark MSR mass from $t\bar{t}$ production cross sections measured as a function of $m_{t\bar{t}}$. Using such a scale setting results in $m_t^{\text{MSRn}}(R = 80\,\text{GeV}) = 169.3^{+0.6}_{-0.7}\,\text{GeV}$, which is in some tension with the result of Ref. [23]. Nonetheless, the value of the top quark MSR mass extracted in this thesis is thus far the most precise value obtained using experimental data.

References

1. Aliev M et al (2011) HATHOR: HAdronic top and heavy quarks crOss section calculatoR. Comput Phys Commun 182:1034–1046
2. Campbell JM, Ellis RK (2010) MCFM for the Tevatron and the LHC. Nucl Phys B Proc Suppl 205-206:10–15 (Blümlein J, Moch S-O, Riemann T (eds))
3. Campbell JM, Ellis RK (2015) Top-quark processes at NLO in production and decay. J Phys G 42:015005
4. Sirunyan AM et al (2020) Running of the top quark mass from proton-proton collisions at $\sqrt{s}$ = 13TeV. Phys Lett B 803:135263
5. Bärnreuther P, Czakon M, Mitov A (2012) Percent-level-preci-sion physics at the tevatron: next- to-next-to-leading order QCD corrections to $q\bar{q}t\bar{t} + X$. Phys Rev Lett 109:132001
6. Czakon M, Mitov A (2012) NNLO corrections to top-pair production at hadron colliders: the all fermionic scattering channels. JHEP 12:054
7. Czakon M, Mitov A (2013) NNLO corrections to top pair production at hadron colliders: the quark gluon reaction. JHEP 01:080
8. Czakon M, Fiedler P, Mitov A (2013) Total top-quark pair-production cross section at hadron colliders through $O(\alpha_S^4)$. Phys Rev Lett 110:252004
9. Kühn JH, Scharf A, Uwer P (2006) Electroweak corrections to top-quark pair production in quark- antiquark annihilation. Eur Phys J C 45:139–150
10. Kühn JH, Scharf A, Uwer P (2007) Electroweak effects in top-quark pair production at hadron colliders. Eur Phys J C 51:37–53
11. Kühn JH, Scharf A, Uwer P (2015) Weak interactions in top-quark pair production at hadron colliders: an update. Phys Rev D 91:014020
12. Lepage GP (1978) A new algorithm for adaptive multidimensional integration. J Comput Tional Phys 27:192–203. ISSN: 0021-9991

13. Lepage GP (1980) VEGAS: an adaptive multidimensional integration program
14. Luscher M (1994) A Portable high quality random number generator for lattice field theory simulations. Comput Phys Commun 79:100–110
15. Van Oldenborgh G (1991) FF: a package to evaluate one loop Feynman diagrams. Comput Phys Commun 66:1–15
16. Alekhin S, Blümlein J, Moch S, Placakyte R (2017) Parton distribution functions, α_S, and heavy quark masses for LHC Run II. Phys Rev D 96:014011
17. Moch S et al (2014) High precision fundamental constants at the TeV scale. arXiv:1405.4781 [hep-ph]
18. Dowling M, Moch S-O (2014) Differential distributions for top-quark hadro-production with a running mass. Eur Phys J C 74:3167
19. Hoang AH, Teubner T (1999) Top quark pair production close to threshold: top mass, width and momentum distribution. Phys Rev D 60:114027
20. Kiyo Y, Kuhn JH, Moch S, Steinhauser M, Uwer P (2009) Top-quark pair production near threshold at LHC. Eur Phys J C 60:375–386
21. Mäkelä T, Hoang A, Lipka K, Moch S-O (2023) Investigation of the scale dependence in the MSR and $\overline{\text{MS}}$ top quark mass schemes for the $\text{t}\bar{\text{t}}$ invariant mass differential cross section using LHC data. arXiv: 2301.03546 [hep-ph]
22. A precise interpretation for the top quark mass parameter in ATLAS monte carlo simulation (2021)
23. Garzelli MV, Kemmler L, Moch S, Zenaiev O (2021) Heavy-flavor hadro-production with heavy quark masses renormalized in the $\overline{\text{MS}}$, MSR and on-shell schemes. JHEP 04:043
24. Alwall J, Herquet M, Maltoni F, Mattelaer O, Stelzer T (2011) MADGRAPH 5: going beyond. JHEP 06:128
25. Alwall J et al (2014) The automated computation of tree-level and next-to-leading order differential cross sections, and their matching to parton shower simulations. JHEP 07:079
26. Bertone V, Frederix R, Frixione S, Rojo J, Sutton M (2014) aMCfast: automation of fast NLO computations for PDF fits. JHEP 08:166
27. Carli T et al (2010) A posteriori inclusion of parton density functions in NLO QCD final-state calcula tions at hadron colliders: the APPLGRID project. Eur Phys J C 66:503–524

Chapter 6
The xFITTER QCD Analysis Framework

This chapter outlines the xFITTER open-source QCD analysis framework [1–4]. The discussion elaborates on the parts of the framework that are used in the analysis presented in Chap. 7, as well as the author's contributions to the framework, developed in the context of the thesis and included in the release version 2.2.0 available at [4]. In particular, the interface for the calculation of SMEFT contributions to the inclusive jet production cross sections, and the production of top quarks as single quarks or $t\bar{t}$ pairs are discussed. Moreover, the MSR renormalization scheme has been introduced for the fixed-order prediction of the $t\bar{t}$ production cross section.

Section 6.1 gives a brief overview of the framework and a discussion of the minimization methods, while the profiling procedure is explained in Sect. 6.2 and the handling of experimental uncertainties in Sect. 6.3. The treatment of heavy quarks is discussed in Sect. 6.4. Section 6.5 describes the determination of PDF parametrizations and the parametrization scan. Section 6.6 details some of the tools and interfaces available for computing theoretical predictions.

6.1 Overview and Minimization Methods

The PDFs have become the primary contribution to the uncertainties in many of the contemporary precision measurements performed at the LHC [5–7]. The xFITTER framework [1–4] is a unique open-source tool for the determination of PDFs and fundamental parameters of the SM, such as the strong coupling constant α_S, quark masses or the EW mixing angle. Moreover, the framework offers built-in plotting tools for visualizing the results.

xFITTER is capable of processing both fixed target and collider experiment data, including measurements from ep, pp and $p\bar{p}$ collisions. Thus, a wide range of Bjorken-x can be probed. The framework contains a wide variety of theoretical options, along with numerous methods and schemes that can be used in the process of PDF

T. Mäkelä, *Towards Global Interpretation of LHC Data*, Springer Theses,
https://doi.org/10.1007/978-3-031-29779-3_6

determination. xFitter supports running quark mass schemes, and has been used for the extraction of the $\overline{\text{MS}}$ charm mass in Ref. [8]. xFitter can also be used for performing SMEFT analyses through an implementation of EFT contributions for 4-quark CI via an interface to CIJET. This way, the SMEFT parameters can be extracted simultaneously with the PDFs and SM parameters. Furthermore, as an increasing amount of PDF analyses are extended to NNLO, it is possible that NLO QED effects become important. In xFitter, the QED effects can be included at NLO. This is demonstrated in the determination of the photon PDF in a NNLO QCD and NLO QED analysis in Ref. [9]. Previously, xFitter has been utilized by the CMS Collaboration in the analyses of the measurements of the double-differential inclusive jet cross sections at $\sqrt{s} = 7\,\text{TeV}$ [10] and $\sqrt{s} = 8\,\text{TeV}$ [11]. In the context of top quark production cross sections, xFitter has been used by the CMS Collaboration in the analyses of the measurement of the normalized triple-differential $\text{t}\bar{\text{t}}$ production cross sections at $\sqrt{s} = 13\,\text{TeV}$ [12] and $\sqrt{s} = 8\,\text{TeV}$ [13], as well as in the double-differential $\text{t}\bar{\text{t}}$ production cross section at $\sqrt{s} = 8\,\text{TeV}$ [14]. Furthermore, the CMS Collaboration has used xFitter in the analyses of Z+charm production [15] and W+charm production [16] at 8 TeV. By the ATLAS Collaboration, xFitter has recently been utilized in the extraction of PDFs from the production of vector bosons in association with jets [17], from a diverse set of ATLAS data measured at $\sqrt{s} = 7, 8$ and 13 TeV [18] and from differential W, Z/γ^* and $\text{t}\bar{\text{t}}$ cross sections [19], as well as for for PDF profiling [20].

The main approach in xFitter is based on the collinear factorization of Eq. (2.55) for hard scattering. The PDFs are convoluted with partonic cross sections, and theory predictions are obtained for the kinematics of the experimental measurements. The predictions are then compared to experimental data, and the PDFs and further parameters are obtained in an iterative χ^2 minimization procedure accounting for all the correlated and uncorrelated uncertainties of the measurements. In the course of the fit, the PDFs are evolved to the scale of the measurement via the DGLAP equations as implemented in the QCDNUM [21] program. Additionally, PDFs can be read and written in the LHAPDF [22, 23] format. Besides the collinear factorization approach, xFitter supports fits of dipole models and transverse-momentum dependent distributions (TMD) [24]. The TMDs can be written and manipulated in the TMDlib format [25, 26], and xFitter has been used for the extraction of TMDs from fits to experimental data in Refs. [27–32]. Further, xFitter has been extended to produce nuclear PDFs [33], as well as for pion PDFs in Ref. [34]. The functionalities of xFitter and the workflow of a QCD analysis are illustrated in Fig. 6.1.

The default algorithm for minimization in xFitter is a modified implementation of the Fortran program MINUIT [35], which has been extended by increasing the number of parameters that can be fitted, as well as allowing for uncertainty analysis with asymmetric error bands in addition to the standard symmetric Hessian treatment. As an alternative, limited support for the CERES [36] minimizer has been added to xFitter.

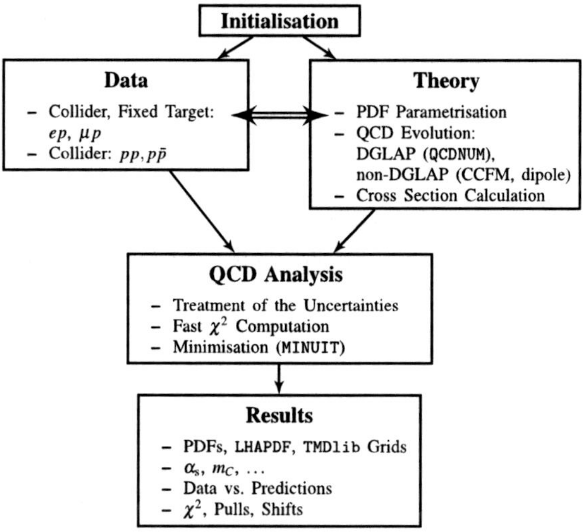

Fig. 6.1 An overview of the XFITTER functionalities and workflow. Image taken from Ref. [1]

Two common variants of the χ^2 representation used in XFITTER are the *covariance matrix* and *nuisance parameter* representations. In the covariance matrix representation, the χ^2 is given by

$$\chi^2 = \sum_{i,j} (m_i - m_i^{\text{data}}) C_{ij}^{-1} (m_j - m_j^{\text{data}}), \tag{6.1}$$

where m_i^{data} are the data points, m_i the corresponding theoretical predictions and C is the covariance matrix composed of the statistical, uncorrelated and systematic contributions such that $C_{ij} = C_{ij}^{\text{stat}} + C_{ij}^{\text{uncorr}} + C_{ij}^{\text{syst}}$. However, the effects of individual sources of uncertainty cannot be distinguished in this representation [1].

The effects of each source of systematic uncertainty can be separated by using the nuisance parameter representation, in which χ^2 is defined as

$$\chi^2(m, b) = \sum_i \frac{m_i^{\text{data}} - m_i \left(1 - \Gamma^{ij} b_j\right)}{\delta_{i,\text{unc}}^2 m_i^2 + \delta_{i,\text{stat}}^2 m_i m_i^{\text{data}} \left(1 - \Gamma^{ij} b_j\right)} + \sum_j b_j^2, \tag{6.2}$$

with $\delta_{i,\text{stat}}$ and $\delta_{i,\text{unc}}$ the relative statistical and uncorrelated systematic uncertainties for the ith measurement, while Γ^{ij} quantifies the measurement's sensitivity to the jth correlated systematic source. The systematic nuisance parameters b_j are determined in the process of the χ^2 minimization, and the effects of different sources can be discriminated.

6.2 The Profiling Procedure

Profiling is a way of assessing the impact of a new set of data on a global PDF set [5, 37–39]. It is based on minimizing a function of the form

$$\chi^2 = \sum_{i=1}^{N_{\text{data}}} \frac{\left(\sigma_i^{\text{exp}} + \Gamma_i^{\alpha,\text{exp}} b_\alpha^{\text{exp}} - \sigma_i^{\text{th}} - \Gamma_i^{\beta,\text{th}} b_\beta^{\text{th}}\right)^2}{\Delta_i^2} + \sum_\alpha (b_\alpha^{\text{exp}})^2 + \sum_\beta (b_\beta^{\text{th}})^2, \quad (6.3)$$

where the nuisance parameter vectors b^{exp} and b^{th} include the correlated experimental and theoretical uncertainties, whose effect on data and theoretical predictions σ^{exp} and σ^{th} is described by the matrices Γ_i^{exp} and Γ_i^{th} for the ith data point out of N_{data}. The indices α and β then run over the experimental and theoretical uncertainty nuisance parameters, respectively. The uncorrelated experimental uncertainties are included in Δ_i. The generalization to asymmetric PDF uncertainties is made by the replacement

$$\Gamma_{i\beta}^{\text{th}} \to \Gamma_{i\beta}^{\text{th}} + \Omega_{i\beta}^{\text{th}} b_\beta^{\text{th}} \quad (6.4)$$

with $\Gamma_{i\beta}^{\text{th}} = (\Gamma_{i\beta}^{\text{th}+} - \Gamma_{i\beta}^{\text{th}-})/2$ and $\Omega_{i\beta}^{\text{th}+} = (\Gamma_{i\beta}^{\text{th}+} + \Gamma_{i\beta}^{\text{th}-})/2$ determined from the predictions' shifts corresponding to the up ($\Gamma_{i\beta}^{\text{th}+}$) and down ($\Gamma_{i\beta}^{\text{th}-}$) eigenvectors [5]. The nuisance parameter values $b^{\text{th(min)}}$ that minimise Eq. (6.3) give the central PDFs f_0' optimized to the profiled dataset in the form

$$f_0' = f_0 + \sum_\beta b_\beta^{\text{th(min)}} \left(\frac{f_\beta^+ - f_\beta^-}{2} - b_\beta^{\text{th(min)}} \frac{f_\beta^+ + f_\beta^- - 2f_0}{2} \right) \quad (6.5)$$

with f_0 the original central PDF and f^+, f^- are the up and down variation eigenvectors [5, 37–39]. For PDFs with asymmetric uncertainties, the covariance matrix C is diagonalized via

$$b^{\text{th}} C b^{\text{th}} = b^{\text{th}} G^T D G b^{\text{th}} = (G' b^{\text{th}})^T G' b^{\text{th}}, \quad (6.6)$$

where D is diagonal and positive definite and G is orthogonal, giving G' the components $G'_{ij} = \sqrt{D_{ik}} G^k{}_j$. Then

$$f_i^{+\prime} = f_0' + \sum_j G'_{ji} \left(\frac{f_j^+ - f_j^-}{2} + G'_{ji} \frac{f_j^+ - 2f_0 + f_j^-}{2} \right),$$
$$f_i^{-\prime} = f_0' - \sum_j G'_{ji} \left(\frac{f_j^+ - f_j^-}{2} - G'_{ji} \frac{f_j^+ - 2f_0 + f_j^-}{2} \right). \quad (6.7)$$

The profiling procedure provides a good and quick method for estimating the change in PDF and non-PDF parameters induced by the addition of new experimental data sets. However, studying the data's full impact on the parametrization and performing simultaneous extractions of the PDF and non-PDF parameters still requires a full PDF fit.

6.3 Treatment of Experimental Uncertainties

The experimental uncertainties are typically propagated to the fitted PDFs by using the Hessian method [40]. Alternatively, this can be done using the method of Monte Carlo (MC) replicas [41, 42].

The statistical and systematic uncertainties of the data are propagated to the PDFs using the Hessian eigenvector method, which makes use of the Hessian matrix defined by the second derivatives of the fitted PDFs at the minimum. After the diagonalization of the Hessian matrix, the eigenvectors corresponding to $\chi^2 = \chi^2_{\min} + 1$ are computed. These correspond to independent sources of uncertainty in the extracted PDFs, since the sets of eigenvectors are within an orthonormal parameter space.

In xFITTER, both symmetric and asymmetric uncertainties are supported. The asymmetric uncertainties $\Delta \mathrm{f}_{\mathrm{exp}^{\pm}}$ for a PDF f are computed as

$$\Delta \mathrm{f}_{\mathrm{exp}^+} = \sqrt{\sum_i^{N_{\mathrm{EV}}} [\max(\mathrm{f}_i^+ - \mathrm{f}_0, \mathrm{f}_i^- - \mathrm{f}_0, 0)]^2}$$

$$\Delta \mathrm{f}_{\mathrm{exp}^-} = \sqrt{\sum_i^{N_{\mathrm{EV}}} [\min(\mathrm{f}_i^+ - \mathrm{f}_0, \mathrm{f}_i^- - \mathrm{f}_0, 0)]^2} \tag{6.8}$$

where f_0 is the central PDF, while f_i^+ and f_i^- represent the results using the up-and down variations of the ith eigenvector, respectively, when the PDF set contains N_{EV} eigenvectors [43].

Alternatively, in the MC method, the fit uncertainties are estimated by creating $\mathcal{O}(> 100)$ pseudo-data replicas by randomly varying the cross section values in the data within their statistical and systematic uncertainties as

$$\sigma_i = \sigma_i \left(1 + \delta_i^{\mathrm{uncorr}} \mathrm{RAND}_i + \sum_j^{N_{\mathrm{syst}}} \delta_{ij}^{\mathrm{uncorr}} \mathrm{RAND}_j\right) \tag{6.9}$$

For each variation, a fit is performed. The central values for the fitted parameters and their uncertainties are estimated as the mean and the root mean square values over the replicas.

6.4 Treatment of Heavy Quarks

The xFITTER framework contains implementations of various schemes for the treatment of the masses of the charm and beauty quarks.[1] In the *zero mass variable flavor number* (ZM-VFN) scheme [44], a heavy quark Q appears as a parton within the proton at scales μ^2 above the production threshold $\sim m_Q^2$, and both the light and heavy quarks are considered massless in the initial and final states of the hard scattering process. The ZM-VFN is to be considered valid only at scales $\mu^2 \gg m_Q^2$, since missing corrections of $\mathcal{O}(m_Q^2/\mu^2)$ make it inaccurate at low μ^2.

The *fixed flavor number* (FFN) is a rigorous QFT scheme [45–47] in which only light quarks and gluons are considered as partons within the proton, while all massive quarks are produced perturbatively in the final state. The number of active flavors in the proton can be fixed to $n_f \in \{3, 4, 5\}$, and are treated as active even at scales beyond the quark production thresholds. The scheme is considered valid at $\mu^2 \approx m_Q^2$. With $n_f = 3$ light flavors, the FFN scheme gives a good description of the experimental data on the production of heavy quarks in DIS [48–50]. However, LHC processes are commonly described with $n_f = 5$, obtained via a decoupling of the heavy quarks and transitioning from 3 flavor PDFs to 4 or 5 flavors, with the prospect of resumming large logarithms in μ^2/m_Q^2 [50].

The aim of the *general mass variable flavor number* (GM-VFN) scheme is to combine the ZM-VFN and FFN schemes. At $\mu^2 \sim m_Q^2$, the heavy quarks are treated according to the FFN scheme, and in the massless scheme at $\mu^2 \gg m_Q^2$. There are several implementations for the interpolation between the two schemes, and the *Thorne-Roberts* GM-VFN [51–53] utilized in this thesis requires the equivalence of the descriptions of the FFN scheme with $n_f = n$ and the GM-VFN scheme with $n_f = n + 1$ above the transition point of the schemes. Moreover, the descriptions must agree for the transition from $n_f = 1$ to a higher number of available quark flavors $n_f = n + 1$.

6.5 PDF Parametrizations

Since the x-shape of the PDFs cannot yet be calculated from first principles, it has to be parametrized at a starting scale Q_0^2. Commonly, the PDFs are parametrized using a polynomial form

$$xf(x) = Ax^B(1-x)^C P(x, D, E), \tag{6.10}$$

where P is a polynomial in x with the coefficients $\{A, B, C, D, E\}$. The exact form of P differs between global PDF collaborations. For example, the HERAPDF parametrization used in this thesis utilizes the ordinary polynomials

[1] Here denoted as heavy quarks.

$$P(x, D, E) = (1 + Dx + Ex^2). \tag{6.11}$$

For the ABMP16 PDF set [54], the collaboration parametrizes the valence quarks at the starting scale with $n_f = 3$ light flavors as

$$xq_v(x) = \frac{2\delta_{q_v\mathrm{u}} + \delta_{q_v\mathrm{d}}}{N_q^v} x^{a_{q_v}} (1-x)^{b_{q_v}} x^{P_{q_v}(x)}, \tag{6.12}$$

where the δ are the Kroenecker symbols defined by $\delta_{qq'} = 1$ if $q = q'$ and $\delta_{qq'} = 0$ otherwise. The parameter N_q^v is determined from the sum rule for fermion number conservation, while a and b are fit parameters. The sea quark distributions are parametrized as

$$xq_s(x) = x\overline{q}_s(x) = A_{q_s}(1-x)^{b_{q_s}} x^{a_{q_s} P_{q_s}(x)}, \tag{6.13}$$

and the gluon PDF as

$$xg(x) = A_g x^{a_g} (1-x)^{b_g} x^{a_g P_g(x)}, \tag{6.14}$$

where the parameter A_g is determined via the momentum conservation sum rule. For a parton $p \in q_v, q_s, g$, the polynomial P_p has the general form

$$P_p(x) = (1 + \gamma_{-1,p} \log(x))(1 + \gamma_{1,p} x + \gamma_{2,p} x^2 + \gamma_{3,p} x^3), \tag{6.15}$$

where the γ are fit parameters.

For the CT14 PDF set [55], the CTEQ Collaboration employs parametrizations of the form

$$xf(x) = x^{a_1} x^{a_2} P(x).$$

Here the polynomial P is written for valence quarks as

$$P_{u_v} = d_0 p_0(y) + d_1 p_1(y) + d_2 p_2(y) + d_3 p_3(y) + d_4 p_4(y),$$

where the d are fit parameters and p are the *Bernstein polynomials*

$$\begin{aligned}
p_0(y) &= (1-y)^4, \\
p_1(y) &= 4y(1-y)^3, \\
p_2(y) &= 6y^2(1-y)^2, \\
p_3(y) &= 4y^3(1-y), \\
p_4(y) &= y^4
\end{aligned}$$

with $y \equiv \sqrt{x}$ to avoid the rapid variations possible in exponential forms. The parameters a_1 and a_2 are set equal for the u_v and d_v distributions. Moreover, $d_4 = 1$ and $d_3 = 1 + a_1/2$. The gluon distribution is parametrized as

$$P_g(y) = g_0[e_0 q_0(y) + e_1 q_1(y) + q_2(y)],$$

with $y = 2\sqrt{x} - x$, and

$$q_0(y) = (1 - y)^2,$$
$$q_1(y) = 2y(1 - y),$$
$$q2(y) = y^2.$$

The parametrizations used by the ABMP and HERAPDF Collaborations are available in xFITTER as default options. Nonetheless, Eq. (6.10) can also be replaced by custom polynomials.

Potentially, any new data set included in a QCD analysis brings additional information on the PDF shapes, which technically means an increase in the number of free parameters in the PDF fit. Therefore, the first step of a QCD analysis is to perform a *parametrization scan*. Using for instance the HERAPDF parametrization of Eq. (6.11), the scan starts with the free parameters A-C, while all D and E set to zero in the first step of the parametrization scan. The additional parameters are included one at a time in the fit, and the one resulting in the best improvement in χ^2 is chosen. The step is repeated, starting from the parametrization now including also the previously chosen parameter. The parameter values and the PDF shapes are monitored at each step to avoid non-physical solutions. The scan terminates after no additional parameter brings an improvement of $\Delta\chi^2 > 1$. The procedure is illustrated in Fig. 6.2.

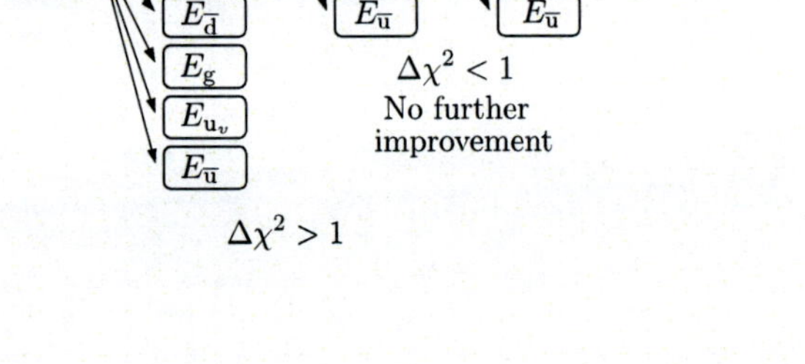

Fig. 6.2 An example of a parametrization scan, starting from a 10-parameter fit with all D and E parameters set to zero. The parameters chosen at each step are highlighted

6.6 Theoretical Predictions in xFITTER

In xFITTER, the theoretical prediction to be used for each dataset is stated as an expression, where each term is computed by objects called *reactions*. Typically, xFITTER forwards the PDFs and relevant theory parameters to a reaction class, which computes its contribution to the cross section and returns it to xFITTER. For instance, the computation can be self-contained in the reaction class, or it may call external dependencies or interpolate the results from precalculated grids. Often a combination of several approaches is required.

6.6.1 *Fast Interpolation Grid Techniques*

The fits performed in a QCD analysis involve multiple recomputations of the cross section. However, the calculation of a full pQCD prediction at NLO or beyond is computationally expensive. This is overcome by the applicaton of *fast interpolation grid techniques*. In this method, an external calculation of the perturbative cross sections is performed only once, and the results are stored into tables of perturbative coefficients that can be represented as a grid in the fractional momenta x_1, x_2 and the QCD scales $\mu_{f,r} = \mu$. This enables high precision predictions to be used quickly without repeating the full NLO calculation at each iteration. Using the tables, the cross section calculation becomes [56]

$$\sigma(\mu) \approx \sum_{n,i,k,l,m} \tilde{\sigma}_{n,i,k,l,m} \left(\alpha_S\left(\mu^{(m)}\right)\right)^n \mathrm{f}_i \left(x_1^{(k)}, x_2^{(l)}, \mu^{(m)}\right), \tag{6.16}$$

where n is the order in α_S, f_i is the PDF for flavor i, and the sums in k, l and m run over the number of points in the corresponding grid. All the information on the perturbative calculation, such as the phase-space restrictions and e.g. the jet definition, is then included in

$$\tilde{\sigma}(\mu) = c_{n,i}(x_1, x_2, \mu) \otimes \left[e^{(k)}(x_1) \cdot e^{(l)}(x_2) \cdot b^{(m)}(\mu)\right], \tag{6.17}$$

where $e^{(k)}$, $e^{(l)}$, $b^{(m)}$ are interpolation functions for the dependencies on x_1, x_2 and μ. Equation (6.17) is independent of the PDFs and α_S; it needs to be evaluated only once, and is then inserted into Eq. (6.16). In the course of the PDF fit, it is hence possible to convolute the coefficients with the PDFs and simply multiply by α_S. The PDFs and α_S are separated from the perturbative coefficients in $\tilde{\sigma}$ by introducing a set of eigenfunctions around a defined number of x-values.

This approach is used in the FASTNLO [57] program interfaced with NLO-Jet++ [58, 59] for jet production calculations in DIS and inclusive single, dijet and three-jet production in hadron-hadron collisions. In the NLOJet++ program, the NLO correction is computed using the Catani-Seymour dipole formalism [60], modified to improve the control of the numerical calculation [58]. The approach is based on cutting the phase space of the dipole subtraction terms [61], and the IR singularities are cancelled directly at the level of the integrand. In the dipole method, a single dipole subtraction term $d\sigma_A$ is defined, reguralizing the real correction in all the singular limits, both soft and collinear. The N-jet cross section is then expressed at NLO in terms of the dipole subtraction and the real and virtual corrections as [61]

$$\sigma^{(1)} = \underbrace{\int_N d\sigma_N^{(1)}}_{\int_N \left(d\sigma_{\text{Virtual}} + \int_1 d\sigma_A\right)} + \underbrace{\int_{N+1} d\sigma_{N+1}^{(1)}}_{\int_{N+1} \left(d\sigma_{\text{Real}} - d\sigma_A\right)}. \tag{6.18}$$

The dipole method does not rely on approximations, and can be implemented numerically in a fully process independent manner, and full control of generating the phase space efficiently is maintained via the exact phase space factorization. Due to preserving Lorentz invariance, changes of frame can be performed via simple momentum transformations. Moreover, there is no need for color ordered subamplitudes, symmetrization, or partial fractioning of the matrix elements, and the use of crossing functions is avoided [58]. In the calculation of the perturbative coefficients $c_{i,n}$, the dependence on μ_f and μ_r is factorized as [57]

$$c_{n,i}(\mu_f, \mu_r) = c^0_{n,i} + \log(\mu_f) c^f_{n,i} + \log(\mu_r) c^r_{n,i}, \tag{6.19}$$

and only the weights c^0, c^r and c^f are stored in scale-independent FASTNLO tables. An interface to FASTNLO then provides xFITTER with the NLO prediction that is the basis of a jet cross section computation. Alternatively, predictions can be obtained at NNLO via the NNLOJet [62–64] program. The processes currently included in NNLOJet are the production of Z, W and H, also with additional jets, single jet production in hadron hadron collisions, dijet production in hadron-hadron and lepton-hadron collisions as well as three-jet production in electron-positron annihilation [62, 63, 65–74]. The NNLO QCD calculations in NNLOJet rely on the antenna subtraction method [75, 76], in which the IR subtraction terms for the double-virtual, real-virtual and double-real contributions are obtained via antenna functions [75–79] describing the unresolved soft and collinear parton radiation from a pair of hard partons, which can be either in the initial or final state. The combinations with final-final, initial-final and initial-initial antennae, each with their respective unresolved phase spaces, are considered [64, 76].

Another tool for fast grid interpolation interfaced to the xFITTER framework is APPLGRID [80]. Its approach makes use of the the sparse nature of the x-dependent weights, allowing for increased flexibility in the choice of scale, since the scale dependence is retained as an additional dimension in the tables. Thus, it enables a posteriori evaluation of the μ_r and μ_f dependence [80]. APPLGRID can be interfaced with programs providing fixed-order calculations for various precesses in pp or $\text{p}\bar{\text{p}}$ collisions, such as the MCFM generator [81]. In the APPLGRID approach, PDFs are represented by storing their values on a two-dimensional grid, and the values are obtained by interpolating between the grid points. To provide good coverage of the full x and μ_f^2 range with uniformly spaced grid points, the variable transformation

$$y_i = \log\left(\frac{1}{x_i}\right) + a(1 - x_i) \quad \text{and} \quad \tau(\mu_f^2) = \log\left(\log\left(\frac{\mu_f^2}{\Lambda^2}\right)\right), \tag{6.20}$$

is used instead of the direct values. Here the index $i \in \{1, 2\}$ is for the incoming hadrons, the parameter a is for increasing the density of points in the high-x region, and the scale Λ should be chosen at $\mathcal{O}(\Lambda_{\mathrm{QCD}})$. A cross section is then computed as

$$\sigma = \sum_p \sum_{l=0}^{n_{\mathrm{sub}}} \sum_{i_{y_1}, i_{y_2}} \sum_{i_\tau} W^{(p)(l)}_{i_{y_1}, i_{y_2}, i_\tau} \left(\frac{\alpha_S\left(\mu_f^{2(i_\tau)}\right)}{4\pi}\right)^p F^{(l)}\left(x_1^{(i_{y_1})}, x_2^{(i_{y_2})}, \mu_f^{2(i_\tau)}\right), \tag{6.21}$$

where n_{sub} is the number of subprocesses contributing to the cross section, the grid weights are given by $W^{(p)(l)}_{i_{y_1}, i_{y_2}, i_\tau}$ and $F^{(l)}$ are the initial state parton combinations. Dependence on the renormalization scale can also be included, and μ_f, μ_r can be varied a posteriori by constant factors [80].

In case full interpolation grids are not available for a given cross section at the desired order, but the calculation is, the corrections to a pQCD cross section prediction σ can also be applied by computing the so-called k-factors

$$k = \frac{\sigma^{\mathrm{Corr}}}{\sigma}. \tag{6.22}$$

Here σ^{Corr} is the value to which the cross section is to be converted to, obtained for instance as the result of an external NNLO computation or a cross section containing corrections for electroweak or nonperturbative effects. The k-factors are computed once for each bin in the data, prior to the fit, and the corrected cross section enters the calculations performed during the fit as

$$\sigma^{\mathrm{Corr}} \approx (\sigma \otimes \mathrm{PDFs}) \cdot k. \tag{6.23}$$

6.6.2 SMEFT Functionality in xFITTER

The CIJET grid interpolation is a modified version of the methods introduced by FASTNLO and APPLGRID. The CIJET interfrace introduces 4-quark CI via the dimension 6 SMEFT operators discussed in Sect. 3.2. The types of CI supported in the interface include purely left-handed, vector-like and axial vector-like CI. In these models, only the Wilson coefficient c_1 multiplying the operator O_1 given in Eq. (3.7) is a free parameter, while c_3 and c_5 multiplying the operators O_3 and O_5 in Eqs. (3.9) and (3.11), respectively, are determined from c_1 as illustrated in Table 6.1. The order

Table 6.1 The dependence of the Wilson coefficients c_3 and c_5 on c_1, which is a free fit parameter in the three specific CI models supported by the xFitter CIJET interface. The coefficients multiply the respective operators O_1, O_3 and O_5 in the SMEFT Lagrangian as indicated in Eq. (3.6)

Type of CI	c_1	c_3	c_5
Purely left-handed:	Free	0	0
Vector-like:	Free	$2c_1$	c_1
Axial-vector-like:	Free	$-2c_1$	c_1

of the computation is taken to be the same as that of the QCD evolution, unless an order higher than NLO is requested, in which case the CI computation reverts to NLO.

To avoid dependence on an external installation of the original CIJET program, the calculations are based on a precomputed CI grid for single jet or dijet cross sections. This is similar to how FASTNLO is used, and the CI contributions provided by CIJET are to be used as an addition to the SM prediction obtained from a FASTNLO grid. The general frame of the CIJET interface is provided in the C++ class `CIJETReader`, while the computations are implemented in the `ReactionCIJET` C++ class invoking the Fortran routines in `cijet.f`.

In each kinematic range, the cross section is calculated as [82]

$$\sigma = \sum_{p,k} \sum_{n_1,n_2,n_3} I_{k,p}(n_1, n_2, n_3, \mu_r/\mu_f) \left((\alpha_S(\mu_r))^p \Phi_k \right) \Big|_{n_1,n_2,n_3}, \tag{6.24}$$

where n_i are grid point indices and the Φ_k are parton luminosities. The relevant parton luminosities Φ_k read [82]

$$\Phi_1 = \sum_{ij=qq',\overline{qq}'} x_1^2 x_2^2 \mathrm{f}_{i,h_1}(x_1, \mu_f) \mathrm{f}_{j,h_2}(x_1, \mu_f) \tag{6.25}$$

$$\Phi_2 = \sum_{ij\in\{qq,\overline{qq}\}} x_1^2 x_2^2 \mathrm{f}_{i,h_1}(x_1, \mu_f) \mathrm{f}_{j,h_2}(x_1, \mu_f) \tag{6.26}$$

$$\Phi_3 = \sum_{ij\in\{q\overline{q}'\}} x_1^2 x_2^2 \mathrm{f}_{i,h_1}(x_1, \mu_f) \mathrm{f}_{j,h_2}(x_1, \mu_f) + [h_1 \leftrightarrow h_2] \tag{6.27}$$

$$\Phi_4 = \sum_{ij\in\{q\overline{q}\}} x_1^2 x_2^2 \mathrm{f}_{i,h_1}(x_1, \mu_f) \mathrm{f}_{j,h_2}(x_1, \mu_f) + [h_1 \leftrightarrow h_2] \tag{6.28}$$

$$\Phi_5 = \sum_{ij\in\{gq,g\overline{q}\}} x_1^2 x_2^2 \mathrm{f}_{i,h_1}(x_1, \mu_f) \mathrm{f}_{j,h_2}(x_1, \mu_f) + [h_1 \leftrightarrow h_2], \tag{6.29}$$

as the gluon-gluon parton luminosity does not contribute to the CI considered here, up to NLO. The grid coefficients I are computed analytically in CIJET. Their values for each kinematic bin are stored in tables allowing for fast interpolations of the bin cross sections for arbitrary choices of the renormalization scale and PDFs [82].

6.6.3 Inclusive Cross Sections for Top Quark Production in xFITTER

The inclusive cross section predictions for top quark production—single t, $\bar{t}$, or $t\bar{t}$ pair, are available in xFITTER via an interface to the HATHOR [83, 84] program. HATHOR includes higher-order perturbative QCD corrections and allows for detailed studies of theoretical uncertainties. The original HATHOR program version 2.0 provides the possibility to compute the total cross section for $t\bar{t}$ production up to NNLO in the pole and $\overline{\text{MS}}$ mass schemes, as well as the single t and $\bar{t}$ production cross sections up to NLO in the pole mass scheme. Provided that the QCD couplings are standard, HATHOR is also applicable to studies involving a hypothetical fourth family of quarks [83].

The HATHOR program was first employed for extracting the mass of the top quark in the $\overline{\text{MS}}$ scheme from $t\bar{t}$ pair production data in hadronic collisions in Refs. [85, 86]. HATHOR has been used by the CMS Collaboration in the extraction of the top quark mass in the $\overline{\text{MS}}$ scheme using $t\bar{t}$ production cross section data measured at $\sqrt{s} = 13$ TeV in Ref. [87], and together with the strong coupling constant in Ref. [88]. Furthermore, predictions for the single top quark production cross section obtained with HATHOR have been utilized in a measurement of the CKM matrix elements in single top quark production in hadronic collisions at $\sqrt{s} = 13$ TeV [89].

In the scope of this thesis, the MSR renormalization scheme is implemented into xFITTER for the interface of the HATHOR code via the reaction HathorMSR, which is a handle to an external installation of the HATHOR program version 2.0. For top quark pair production, support is included for the pole, $\overline{\text{MS}}$, MSRn and MSRp schemes, with the mass evolution discussed in Sects. 2.6.2–2.6.3. The computation of the inclusive cross section, described in Sect. 5.1, is performed within the interface. Therefore, a standard version of HATHOR 2.0, without an internal implementation for the MSR schemes, is sufficient. When using the MSR schemes, the top quark mass can be input to the interface either directly as the MSR mass, or as the $\overline{\text{MS}}$ mass $\overline{m}_t(\overline{m}_t)$, which is then evolved to the corresponding MSR mass at the given scale R.

The predictions for the inclusive single top production cross section are obtained in xFITTER by using the reaction HathorSingleTop, which calls an external installation of HATHOR 2.0. Whereas the standard HATHOR program computes the single top cross sections in the pole mass scheme up to NLO as detailed in Ref. [84], the reaction interface also includes an implementation of the $\overline{\text{MS}}$ scheme.

In the original HATHOR 2.0 program, the calculation of single top production cross sections is based on the MCFM implementation [90–92] and the formulae in [93]. The xFITTER interface retrieves the results from the external program and, if requested, translated numerically to the $\overline{\text{MS}}$ scheme. In this approach, the full NLO cross section in the $\overline{\text{MS}}$ scheme is written in terms of the LO ($\sigma^{(0)}$) and NLO ($\sigma^{(1)}$) contributions as

$$\sigma = (a_S(\overline{m}_t))^2\,\sigma^{(0)}(\overline{m}_t) + (a_S(\overline{m}_t))^3 \left(\sigma^{(1)}(\overline{m}_t) + \overline{m}_t d_1^{\overline{\mathrm{MS}}} \frac{d\sigma^{(0)}}{dm_t}\bigg|_{m_t=\overline{m}_t,\mu_r=\overline{m}_t} \right), \tag{6.30}$$

where a_S is defined in Eq. (2.27), $d_1^{\overline{\mathrm{MS}}}$ is given in Eq. (2.34) and $\overline{m}_t$ is the top quark mass in the $\overline{\mathrm{MS}}$ scheme, all evaluated at $\mu_r = \overline{m}_t$. The dependence on a general μ_r is restored via the expansion in Eq. (2.30).

Since single top production is directly sensitive to the V_{tb} component of the CKM matrix, the reaction interface forwards the values of the CKM matrix elements from xFitter to Hathor, instead of using the standard values of the external Hathor installation. Similarly, the conversion factor hc^2, where h is Planck's constant and c the speed of light, and the EW coupling parameters are forwarded from xFitter to Hathor.

As the extension of the single top quark production cross section to NNLO is currently an on-going effort, all scale-dependent NNLO terms are already implemented into Hathor for possible future development. The NNLO vertex corrections and related real corrections are calculated for the t-channel in Ref. [94], excluding double-box topologies. The reduction to master integrals is presented in Ref. [95] for the full set of corrections at the 2-loop level. Recent results for the t-channel are reported in Ref. [96] and for the s-channel in [97]. Therefore, an NNLO formula with the form of Eq. (5.6) is also implemented into xFitter, and can be enabled easily once the pole mass scheme predictions are implemented into a future public version of Hathor.

References

1. Alekhin S et al (2015) HERAFitter, open source QCD fit project. Eur Phys J C 75:304
2. Bertone V, Botje M, Britzger D et al (2018) XFITTER 2.0.0: an open source QCD fit framework. PoS(DIS2017)203
3. Abdolmaleki H et al (2022) xFitter: An Open Source QCD Analysis Framework. A resource and refer ence document for the Snowmass study. arXiv: 2206.12465 [hep-ph]
4. XFitter developers' team. https://www.xfitter.org/xFitter/
5. Camarda S et al (2015) QCD analysis of W- and Z-boson production at Tevatron. Eur Phys J C 75:458
6. Accomando E et al (2019) PDF profiling using the forward-backward asymmetry in Neutral Current Drell-Yan production. JHEP 10:176
7. Amoroso S et al (2021) Longitudinal Z-boson polarization and the Higgs boson production cross section at the Large Hadron Collider. Phys Lett B 821:136613
8. Bertone V et al (2016) A determination of mc(mc) from HERA data using a matched heavy-flavor scheme. JHEP 08:050
9. Giuli F et al (2017) The photon PDF from high-mass Drell-Yan data at the LHC. Eur Phys J C 77:400
10. Khachatryan V et al (2015) Constraints on parton distribution functions and extraction of the strong coupling constant from the inclusive jet cross section in pp collisions at $\sqrt{s}$ = 7 TeV. Eur Phys J C 75:288

11. Khachatryan V et al (2017) Measurement and QCD analysis of double-differen-tial inclusive jet cross sections in pp collisions at $\sqrt{s}$ = 8 TeV and cross section ratios to 2.76 and 7 TeV. JHEP 03:156
12. Sirunyan AM et al (2020) Measurement of tt normalised multi-differential cross sections in pp collisions at $\sqrt{s}$ =13 TeV, and simultaneous determination of the strong coupling strength, top quark pole mass, and parton distribution functions. Eur Phys J C 80:658
13. Sirunyan AM et al (2017) Measurement of the triple-differential dijet cross section in proton-proton collisions at $\sqrt{s}$ = 8 TeV and constraints on parton distribution functions. Eur Phys J C 77:746
14. Sirunyan AM et al (2017) Measurement of double-differential cross sections for top quark pair pro duction in pp collisions at $\sqrt{s}$ = 8 TeV and impact on parton distribution functions. Eur Phys J C 77:459
15. Sirunyan AM et al (2018) Measurement of associated Z + charm production in proton-proton collisions at $\sqrt{s}$ = 8 TeV. Eur Phys J C 78:287
16. Tumasyan A et al (2021) Measurements of the associated production of a W boson and a charm quark in proton-proton collisions at $\sqrt{s}$ = 8 TeV. arXiv: 2112.00895 [hep-ex]
17. Aad G et al (2021) Determination of the parton distribution functions of the proton from ATLAS measurements of differential $W^{\pm}$ and Z boson production in association with jets. JHEP 07:223
18. Aad G et al (2022) Determination of the parton distribution functions of the proton using diverse ATLAS data from pp collisions at $\sqrt{s}$ = 7, 8 and 13 TeV. Eur Phys J C 82:438
19. Determination of the parton distribution functions of the proton from ATLAS measurements of differential W and Z/γ^* and $t\bar{t}$ cross sections (2018)
20. Aaboud M et al (2017) Precision measurement and interpretation of inclusive W+, W^- and Z/γ^* production cross sections with the ATLAS detector. Eur Phys J C 77:367
21. Botje M (2011) QCDNUM: fast QCD evolution and convolution. Comput Phys Commun 182:490
22. Buckley A et al (2015) LHAPDF6: parton density access in the LHC precision era. Eur Phys J C 75:132
23. Whalley MR, Bourilkov D, Group RC (2005) The Les houches accord PDFs (LHAPDF) and LHAGLUE. In: HERA and the LHC: A Workshop on the Implications of HERA and LHC Physics (Startup Meeting, CERN, 26–27 March 2004; Midterm Meeting, CERN, 11–13 October 2004). pp 575–581. arXiv: hep-ph/0508110
24. Angeles-Martinez R et al (2015) Transverse momentum dependent (TMD) parton distribution functions: status and prospects. Acta Phys Polon B 46:2501–2534
25. Hautmann F et al (2014) TMDlib and TMDplotter: library and plotting tools for transverse-momentum dependent parton distributions. Eur Phys J C 74:3220
26. Abdulov NA et al (2021) TMDlib2 and TMDplotter: a platform for 3D hadron structure studies. Eur Phys J C 81:752
27. Hautmann F, Jung H (2014) Transverse momentum dependent gluon density from DIS precision data. Nucl Phys B 883:1–19
28. Dooling S, Hautmann F, Jung H (2014) Hadroproduction of electroweak gauge boson plus jets and TMD parton density functions. Phys Lett B 736:293–298
29. Hautmann F, Jung H, Monfared ST (2014) The CCFM uPDF evolution uPDFevolv Version 1.0.00. Eur Phys J C 74:3082
30. Bermudez Martinez A et al (2019) Collinear and TMD parton densities from fits to precision DIS measurements in the parton branching method. Phys Rev D 99:074008
31. Hautmann F, Jung H, Lelek A, Radescu V, Zlebcik R (2018) Collinear and TMD quark and gluon densities from parton branching solution of QCD evolution equations. JHEP 01:070
32. Hautmann F, Jung H, Lelek A, Radescu V, Zlebcik R (2017) Soft-gluon resolution scale in QCD evolution equations. Phys Lett B 772:446–451
33. Walt M, Helenius I, Vogelsang W (2019) Open-source QCD analysis of nuclear parton distribution functions at NLO and NNLO. Phys Rev D 100:096015
34. Novikov I et al (2020) Parton distribution functions of the charged pion within the xFitter frame work. Phys Rev D 102:014040

35. James F, Roos M (1975) Minuit-a system for function minimization and analysis of the param eter errors and correlations. Comput Phys Commun 10:343–367. ISSN: 0010-4655
36. Agarwal S, Mierle K (2012) Ceres solver: tutorial & reference google Inc
37. Paukkunen H, Zurita P (2014) PDF reweighting in the Hessian matrix approach. JHEP 12:100
38. Schmidt C, Pumplin J, Yuan CP, Yuan P (2018) Updating and optimizing error parton distribution function sets in the Hessian approach. Phys Rev D 98:094005
39. Abdul Khalek R, Bailey S, Gao J, Harland-Lang L, Rojo J (2018) Towards ultimate parton distributions at the high-luminosity LHC. Eur Phys J C 78:962
40. Pumplin J et al (2001) Uncertainties of predictions from parton distribution functions. II. The Hessian method. Phys Rev D 65:014013
41. Giele WT, Keller S (1998) Implications of hadron collider observables on parton distribution function uncertainties. Phys Rev D 58:094023
42. Giele WT, Keller SA, Kosower DA (2001) Parton distribution function uncertainties
43. Rabbertz K (2017) Jet physics at the LHC. ISBN: 978-3-319-42113-1. https://doi.org/10.1007/978-3-319-42115-5
44. Collins JC, Tung W-K (1986) Calculating heavy quark distributions. Nucl Phys B 278:934–950. ISSN: 0550-3213
45. Laenen E, Riemersma S, Smith J, van Neerven W (1993) Complete O(α_S) corrections to heavy flavour structure functions in electroproduction. Nucl Phys B 392:162–228. ISSN: 0550-3213
46. Laenen E, Riemersma S, Smith J, van Neerven W (1992) On the heavy-quark content of the nucleon. Phys Lett B 291:325–328. ISSN: 0370-2693
47. Riemersma S, Smith J, van Neerven W (1995) Rates for inclusive deep-inelastic electroproduction of charm quarks at HERA. Phys Lett B 347:143–151. ISSN: 0370-2693
48. Abramowicz H et al (2018) Combination and QCD analysis of charm and beauty production cross section measurements in deep inelastic ep scattering at HERA. Eur Phys J C 78:473
49. Accardi A et al (2016) A critical appraisal and evaluation of modern PDFs. Eur Phys J C 76:471
50. Alekhin S, Blumlein J, Moch S (2020) Heavy-flavor PDF evolution and variable-flavor number scheme uncertainties in deep-inelastic scattering. Phys Rev D 102:054014
51. Thorne RS, Roberts RG (1998) An Ordered analysis of heavy flavor production in deep inelastic scattering. Phys Rev D 57:6871
52. Thorne RS (2006) A variable-flavor number scheme for NNLO. Phys Rev D 73:054019
53. Thorne RS (2012) Effect of changes of variable flavor number scheme on parton distribution functions and predicted cross sections. Phys Rev D 86:074017
54. Alekhin S, Blumlein J, Moch S, Placakyte R (2017) Parton distribution functions, α_S, and heavy quark masses for LHC Run II. Phys Rev D 96:014011
55. Dulat S et al (2016) New parton distribution functions from a global analysis of quantum chromodynamics. Phys Rev D 93:033006
56. Kluge T, Rabbertz K, Wobisch M (2006) FastNLO: fast pQCD calculations for PDF fits. In: 14th International Workshop on Deep Inelastic Scattering. pp 483–486. https://doi.org/10.1142/9789812706706_0110. arXiv: hep-ph/0609285
57. Britzger D, Rabbertz K, Stober F, Wobisch M (2012) New features in version 2 of the FASTNLO project. In: 20th International Workshop on Deep-Inelastic Scattering and Related Subjects. p 217. https://doi.org/10.3204/DESY-PROC-2012-02/165. arXiv: 1208.3641 [hep-ph]
58. Nagy Z (2002) Three jet cross-sections in hadron-hadron collisions at next-to-leading order. Phys Rev Lett 88:122003
59. Nagy Z (2003) Next-to-leading order calculation of three jet observables in hadron-hadron collision. Phys Rev D 68:094002
60. Catani S, Seymour MH (1997) A General algorithm for calculating jet cross-sections in NLO QCD. Nucl Phys B 485:291–419 [Erratum: Nucl Phys B 510:503–504 (1998)]
61. Nagy Z, Trocsanyi Z (1999) Next-to-leading order calculation of four jet observables in electron positron annihilation. Phys Rev D 59:014020 [Erratum: Phys Rev D 62:099902 (2000)]
62. Currie J, Glover EWN, Pires J (2017) Next-to-next-to leading order QCD predictions for single jet inclusive production at the LHC. Phys Rev Lett 118:072002

63. Currie J et al (2018) Infrared sensitivity of single jet inclusive production at hadron colliders. JHEP 10:155
64. Gehrmann T et al (2018) Jet cross sections and transverse momentum distributions with NNLOJET. PoS(RADCOR2017)074 (Hoang A, Schneider C (eds))
65. Gehrmann-De Ridder A, Gehrmann T, Glover EWN, Huss A, Morgan TA (2016) Precise QCD predictions for the production of a Z boson in association with a hadronic jet. Phys Rev Lett 117:022001
66. Gehrmann-De Ridder A, Gehrmann T, Glover EWN, Huss A, Morgan TA (2016) The NNLO QCD corrections to Z boson production at large transverse momentum. JHEP 07:133
67. Gehrmann-De Ridder A, Gehrmann T, Glover EWN, Huss A, Morgan TA (2016) NNLO QCD corrections for Drell-Yan p Z T and ø* observables at the LHC. JHEP 11:094 [Erratum: JHEP 10:126 (2018)]
68. Gauld R, Gehrmann-De Ridder A, Gehrmann T, Glover EWN, Huss A (2017) Precise predictions for the angular coefficients in Z-boson production at the LHC. JHEP 11:003
69. Gehrmann-De Ridder A, Gehrmann T, Glover EWN, Huss A, Walker DM (2018) Next to-next-to-leading-order QCD corrections to the transverse momentum distribution of weak Gauge Bosons. Phys Rev Lett 120:122001
70. Chen X, Cruz-Martinez J, Gehrmann T, Glover EWN, Jaquier M (2016) NNLO QCD corrections to Higgs boson production at large transverse momentum. JHEP 10:066
71. Currie J et al (2017) Precise predictions for dijet production at the LHC. Phys Rev Lett 119:152001
72. Currie J, Gehrmann T, Niehues J (2016) Precise QCD predictions for the production of dijet final states in deep inelastic scattering. Phys Rev Lett 117:042001
73. Currie J, Gehrmann T, Huss A, Niehues J (2017) NNLO QCD corrections to jet production in deep inelastic scattering. JHEP 07: 018 [Erratum: JHEP 12:042 (2020)]
74. Gehrmann T, Glover EWN, Huss A, Niehues J, Zhang H (2017) NNLO QCD corrections to event orientation in e +e^{-} annihilation. Phys Lett B 775:185–189
75. Gehrmann-De Ridder A, Gehrmann T, Glover EWN (2005) Quark-gluon antenna functions from neutralino decay. Phys Lett B 612:36–48
76. Currie J, Glover EWN, Wells S (2013) Infrared structure at NNLO using antenna subtraction. JHEP 04:066
77. Daleo A, Gehrmann T, Maitre D (2007) Antenna subtraction with hadronic initial states. JHEP 04:016
78. Daleo A, Gehrmann-De Ridder A, Gehrmann T, Luisoni G (2010) Antenna subtraction at NNLO with hadronic initial states: initial-final configurations. JHEP 01:118
79. Gehrmann T, Monni PF (2011) Antenna subtraction at NNLO with hadronic initial states: real virtual initial-initial configurations. JHEP 12:049
80. Carli T et al (2010) A posteriori inclusion of parton density functions in NLO QCD final-state calculations at hadron colliders: the APPLGRID project. Eur Phys J C 66:503–524
81. Campbell JM, Ellis RK (1999) An update on vector boson pair production at hadron colliders. Phys Rev D 60:113006
82. Gao J (2013) CIJET: a program for computation of jet cross sections induced by quark contact inter actions at hadron colliders. Comput Phys Commun 184:2362
83. Aliev M et al (2011) HATHOR: HAdronic top and heavy quarks crOss section calculatoR. Comput Phys Commun 182:1034–1046
84. Kant P et al (2015) HatHor for single top-quark production: updated predictions and uncertainty estimates for single top-quark production in hadronic collisions. Comput Phys Commun 191:74–89
85. Langenfeld U, Moch S, Uwer P (2009) Measuring the running top-quark mass. Phys Rev D 80:054009
86. Moch S, Langenfeld U, Uwer P (2010) The Top-quark's running mass. PoS(RADCOR2009)030
87. Sirunyan AM et al (2020) Running of the top quark mass from proton-proton collisions at $\sqrt{s}$ = 13TeV. Phys Lett B 803:135263

88. Sirunyan AM et al (2019) Measurement of the tt production cross section, the top quark mass, and the strong coupling constant using dilepton events in pp collisions at $\sqrt{s} = 13$ TeV. Eur Phys J C 79:368
89. Sirunyan AM et al (2020) Measurement of CKM matrix elements in single top quark t-channel pro duction in proton-proton collisions at $\sqrt{s} = 13$ TeV. Phys Lett B 808:135609
90. Campbell JM, Frederix R, Maltoni F, Tramontano F (2009) Next-to-leading-order predictions for t-channel single-top production at hadron colliders. Phys Rev Lett 102:182003
91. Campbell JM, Tramontano F (2005) Next-to-leading order corrections to Wt production and decay. Nucl Phys B 726:109–130
92. Campbell JM, Ellis RK, Tramontano F (2004) Single top production and decay at next-to-leading order. Phys Rev D 70:094012
93. Harris BW, Laenen E, Phaf L, Sullivan Z, Weinzierl S (2002) The fully differential single top quark cross-section in next to leading order QCD. Phys Rev D 66:054024
94. Brucherseifer M, Caola F, Melnikov K (2014) On the NNLO QCD corrections to single-top production at the LHC. Phys Lett B 736:58–63. ISSN: 0370-2693
95. Assadsolimani M, Kant P, Tausk B, Uwer P (2014) Calculation of two-loop QCD corrections for hadronic single top-quark production in the t channel. Phys Rev D 90:114024
96. Campbell J, Neumann T, Sullivan Z (2021) Single-top-quark production in the t-channel at NNLO. JHEP 02:040
97. Liu ZL, Gao J (2018) S-channel single top quark production and decay at next-to-next-to-leading order in QCD. Phys Rev D 98:071501

Chapter 7
Interpretation of CMS Measurements in Terms of SM Parameters and Search for New Physics

This chapter discusses the SM QCD and SMEFT analyses of the CMS inclusive jet and normalized triple-differential $t\bar{t}$ production cross section data at $\sqrt{s} = 13$ TeV, which are used together with the inclusive DIS cross section measurements at HERA. Furthermore, the impact of the CMS data on a global PDF set is studied. The studies are performed in the context of this thesis, and also presented in Ref. [1].

The experimental data sets used in the analysis are listed in Sect. 7.1. Details on the general strategy of the QCD analysis are given in Sect. 7.2, while Sect. 7.3 discusses the invoked theoretical predictions. Section 7.4 details a full PDF fit using SM predictions at NNLO, where $\alpha_S(m_Z)$ is determined simultaneously with the PDFs. Section 7.5 discusses the SMEFT interpretation at NLO in terms of a simultaneous extraction of the PDFs, $\alpha_S(m_Z)$, m_t^{pole} and the Wilson coefficient c_1, $\alpha_S(m_Z)$ and m_t^{pole}; additionally, an alternative analysis performed using SM predictions is presented. Section 7.6 examines the effect of the CMS 13 TeV data on a global PDF set via the profiling technique [2–4] implemented in xFITTER, using the CT14 [5] PDFs at NLO and NNLO.

7.1 Experimental Data

In order to proceed towards a global interpretation of the LHC data, the analysis presented in this thesis includes a simultaneous extraction of the PDFs and SM parameters, with the addition of setting constraints on new physics. Ensuring that all parameters are properly constrained requires a diverse set of experimental data, and the selected measurements are outlined below.

As discussed in Sect. 2.8.1, the basis of PDF fits are the CC and NC deep inelastic scattering (DIS) cross section measurements in $e^{\pm}p$ collisions at HERA [6]. The data are sensitive to light quarks at low and medium x. Additionally, they are sensitive to

T. Mäkelä, *Towards Global Interpretation of LHC Data*, Springer Theses,
https://doi.org/10.1007/978-3-031-29779-3_7

the gluon PDF via scaling violations, and the measurement of the NC cross section also constrains the sea quark distribution at low x. The utilized dataset is a combination of all inclusive DIS cross sections measured by the H1 and ZEUS Collaborations during the full runtime of HERA; therefore, results at different centre-of-mass energies are included. The consistency of the results is ensured by extensive evaluations of the correlated and uncorrelated systematic uncertainties [6]. The NC cross sections are determined in the kinematic range of $6 \cdot 10^{-7} \leq x \leq 0.65$ and the CC cross sections in $1.3 \cdot 10^{-2} \leq x \leq 0.40$.

Although the HERA data give the main constraints on the PDFs, additional data sets give information of the behavior of the PDFs in kinematic regions not covered by the HERA data. The HERA data are used together with the CMS double-differential inclusive jet production cross section at $\sqrt{s} = 13$ TeV, corresponding to an integrated luminosity of 33.5 fb^{-1}. The data are shown in Fig. 7.1, and are sensitive to the high-x gluon PDF and $\alpha_S(m_Z)$. The jets are reconstructed using the anti-k_T jet algorithm [7] with size parameter $\mathcal{R} = 0.7$ due to suffering from smaller out-of-cone radiation effects, and since the theoretical predictions describe the experimental data better than with smaller $\mathcal{R}$ values [1]. It is also argued in Refs. [8–10] that using $\mathcal{R} = 0.7$ instead of $\mathcal{R} = 0.4$ enhances the stability of the NNLO calculations. Furthermore, the CMS normalized triple-differential $t\bar{t}$ cross section at $\sqrt{s} = 13$ TeV [11], corresponding to an integrated luminosity of 35.9 fb^{-1}, are used in the analysis performed at NLO. The variables for the triple-differential cross sections are chosen in order to enhance sensitivity to the PDFs, α_S, and m_t. As demonstrated in [12], the combination of the variables $y_{t\bar{t}}$ and $m_{t\bar{t}}$ provides sensitivity to the PDFs, while the N_{jet} distribution provides sensitivity to α_S and the $m_{t\bar{t}}$ distribution to m_t. As seen in Ref. [11], the top quark pole mass m_t^{pole}, the gluon PDF, and $\alpha_S(m_Z)$ are closely correlated in the triple-

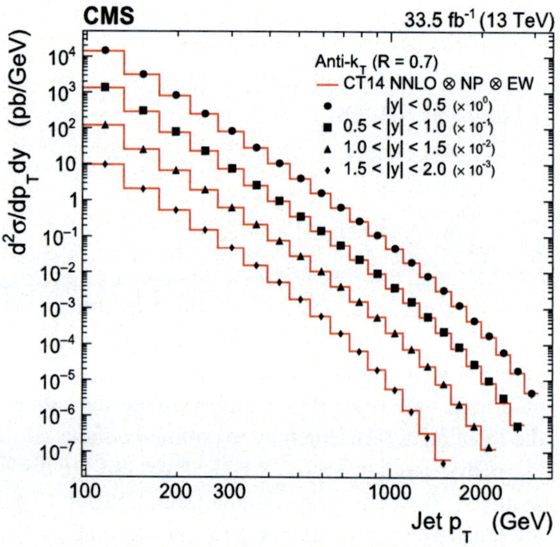

Fig. 7.1 The CMS measurement of the inclusive jet cross section in pp collisions at $\sqrt{s} = 13$ TeV, compared to theoretical predictions using the CT14 PDF. Nonperturbative and electroweak corrections discussed in Sect. 2.7 have been accounted for. Different rapidity regions have been scaled with constant factors for clarity [1]

differential cross section for $t\bar{t}$ production, hence providing additional sensitivity to the gluon distribution and to $\alpha_S(m_Z)$. Combining these data allows the simultaneous extraction of the proton PDFs and the values of $\alpha_S(m_Z)$ and m_t^{pole}. Although inclusive jet production is not sensitive to m_t^{pole}, it has strong constraints on the gluon PDF and $\alpha_S(m_Z)$, which is reflected in the improved value and uncertainty of m_t^{pole} in a fit that utilized both $t\bar{t}$ and inclusive jet production cross section data. Using the LO kinematic relation $x = m_{t\bar{t}}/\sqrt{s}\exp(\pm y_{t\bar{t}})$, the normalized triple-differential $t\bar{t}$ measurement at $\sqrt{s} = 13$ TeV is expected to be sensitive to x values within $0.01 < x < 0.1$ [11].

Correlations of the sources of experimental statistical and systematic uncertainty are included for each data set. The HERA DIS measurements and the CMS data are taken to be uncorrelated. The common systematic uncertainties associated with

Table 7.1 The common uncertainties that are treated as fully correlated in the CMS $t\bar{t}$ and inclusive jet cross section measurements

Name	Description
AbsoluteMPFBias, AbsoluteScale, AbsoluteStat	Flat absolute scale uncertainties. Main uncertainties arise from the combined photon, $Z \to ee$ (electron) and $Z \to \mu\mu$ (tracking) reference scale (AbsoluteScale), and the correction for FSR + ISR (MPFBias)
FlavorQCD	Jet flavor based on differences in uds/c/b-quark and gluon responses between PYTHIA 6 Z2 [13] and HERWIG++ 2.3
Fragmentation	High-p_T extrapolation based on the differences in the description of fragmentation and the underlying event in PYTHIA 6 Z2 and HERWIG++ 2.3
PileUpDataMC	Uncertainty on the data / MC scale factor for offset correction
PileUpPtBB, PileUpPtEC1, PileUpPtEC2, PileUpPtHF, PileUpPtRef	The dependence of the pile-up offset on jet p_T is estimated from matched MC, with and without PU overlay. The uncertainty is essentially obtained from the difference of PU inside and outside of the jets, propagated through the L2 (BB,EC,HF) and L3 (Ref) data-driven methods
RelativeBal	Full difference between log-linear fits of the missing transverse energy projection fraction (MPF) and p_T-balance methods
RelativeFSR	η-dependence uncertainty due to correction for ISR and FSR, estimated from difference between the MPF log-linear residual L2Res correction from PYTHIA 8 and HERWIG++, after each has been corrected for their own ISR+FSR correction

Table 7.2 The common uncertainties that are treated as fully correlated in the CMS $t\bar{t}$ and inclusive jet cross section measurements

Name	Description
RelativeJEREC1, RelativeJEREC2, RelativeJERHF	η-dependence uncertainty from jet p_T resolution (JER). Derived by varying the JER scale factors up and down within uncertainties. The JER uncertainties are assumed fully correlated for the endcap within tracking (EC1), endcap outside tracking (EC2) and hadronic forward (HF) regions
RelativePtBB, RelativePtEC1, RelativePtEC2, RelativePtHF	The half-difference between the MPF method log-linear and constant fits versus p_T
RelativeStatEC, RelativeStatFSR, RelativeStatHF	Statistical uncertainties in the determination of η-dependence, calculated from the error matrix of the final state radiation correction fit versus η, or log-linear L2Res fit versus p_T. The latter only matters in the EC and HF, where the uncertainty is provided as correlated over these wide regions
SinglePionECAL, SinglePionHCAL	High-p_T extrapolation based on propagating variations of $\pm3\%$ in the single particle response in ECAL and HCAL to PF Jets
TimePtEta	Arises from the time dependence of jet energy calibration (JEC) between run epochs. Taken as a difference between the luminosity-weighted average of the corrections per epoch, and a single correction derived from the full sample. The uncertainties for individual epochs are calculated as a weighted difference to the full sample, so that their luminosity-weighted quadratic sum equals TimePtEta

the jet energy scale are treated as fully correlated in the CMS $t\bar{t}$ and inclusive jet cross section measurements. These uncertainties are listed along with their brief descriptions in Tables 7.1 and 7.2.

7.2 The General Strategy

The QCD analysis is performed using the open-source framework xFitter [14–17] version 2.2.1, extended with an interface to the CIJET [18] program for SMEFT predictions, as described in Chap. 6. The DGLAP equations are evolved with QCDNUM [19] version 17-01/14. In different parts of the analysis, the evolution is performed either at NLO or NNLO, corresponding to the order of the available theoretical predictions. The charm and bottom quark pole masses are set to $m_c = 1.47$ GeV and

$m_b = 4.5$ GeV, respectively. Their contributions are treated in the Thorne–Roberts [20–22] variable-flavor number scheme at NLO. The m_t^{pole} and $\alpha_S(m_Z)$ are free parameters in the PDF fits. The DIS data are restricted to high Q^2 by the requirement $Q^2_{\text{min}} = 7.5\,\text{GeV}^2$.

The PDF determination follows the HERAPDF [6, 23] approach. The parametrized PDFs are the gluon distribution $xg(x)$, the valence quark distributions $xu_v(x)$ and $xd_v(x)$, as well as $x\overline{U}(x)$ for the up- and $x\overline{D}(x)$ for the down-type antiquarks, respectively. At the starting scale of QCD evolution $Q_0^2 = 1.9\,\text{GeV}^2$, the general form of the parametrization of a PDF f is given in Eqs. (6.10)–(6.11), with the normalization parameters A_{u_v}, A_{d_v} and A_g determined by the QCD sum rules. The small-x behaviour of the PDFs is driven by the B parameters, whereas the C parameters are responsible for the shape of the distribution as $x \to 1$. The relations

$$\begin{cases} x\overline{U}(x) = x\overline{u}(x), \\ x\overline{D}(x) = x\overline{d}(x) + x\overline{s}(x) \end{cases} \tag{7.1}$$

are assumed for the up, down, and strange antiquarks $x\overline{u}(x)$, $x\overline{d}(x)$, and $x\overline{s}(x)$. The so-called singlet, or the sea quark distribution, is defined by the relation

$$x\Sigma(x) = 2x\overline{u}(x) + x\overline{d}(x) + x\overline{s}(x). \tag{7.2}$$

Following Ref. [6], $x\overline{u}$ and $x\overline{d}$ are constrained to have the same normalization in the $x \to 0$ limit by the requirements

$$\begin{cases} B_{\overline{U}} = B_{\overline{D}}, \\ A_{\overline{U}} = A_{\overline{D}}(1 - f_s). \end{cases} \tag{7.3}$$

Following Ref. [6], the strangeness fraction

$$f_s = \frac{\overline{s}}{\overline{d} + \overline{s}} \tag{7.4}$$

is fixed to $f_s = 0.4$.

The D_f and E_f parameters probe the sensitivity of the results to the specific selected functional form. The form of the parametrization is determined using the procedure described in Sect. 6.5.

7.2.1 Contributions to the Uncertainties

The uncertainties of the PDFs and non-PDF parameters extracted in the full QCD fits are estimated similarly to the procedure of HERAPDF [6, 23], which accounts for fit, model, and parametrization uncertainties.

Table 7.3 Upper and lower limits for the variations used in the determination of model uncertainties. The heavy quark masses correspond to the pole masses

Parameter	Lower limit	Upper limit
m_b	4.25 GeV	4.75 GeV
m_c	1.41 GeV	1.53 GeV
Q^2_{min}	5.0 GeV^2	10.0 GeV^2
f_s	0.32	0.48
Q^2_0	1.7 GeV^2	2.1 GeV^2

The **Fit uncertainties** originate from the sources of uncertainty in the experimental data. They are obtained via the Hessian method [24] explained in Sect. 6.3, with the tolerance criterion $\Delta\chi^2 = 1$ implied, corresponding to the 68% confidence level (CL). For comparison, the fit uncertainties are estimated also with the alternative Monte Carlo (MC) approach explained in Sect. 6.3.

The **Parametrization uncertainty** is estimated by independently adding further D and E parameters to the functional forms of all PDFs, one at a time. An envelope of the maximal differences between the central fit and all parametrization variations is constructed and taken as the resulting uncertainty.

The **Model uncertainties** arise from varying the values assumed for the heavy quark masses m_b and m_c, the strangeness fraction f_s, the Q^2_{min} imposed on the HERA data, and the starting scale Q^2_0, within the limits given in Table 7.3. Furthermore, the **scale uncertainty** is considered as a part of the model uncertainty in the fits. It is obtained by varying the renormalization and factorization scales up and down independently by a factor of 2, and cases with $\mu_f/\mu_r = 4^{\pm 1}$ are excluded. The fit is repeated for every variation, and the one with the largest difference to the central result is included as the scale uncertainty. Finally, the model uncertainty is attained as the quadratic sum of the contributions of all model variations.

The **total uncertainty** of the fits is acquired by adding the fit and model uncertainties in quadrature, and the parametrization uncertainties linearly.

7.3 Theoretical Predictions in the QCD Analysis

The basic ingredient of the theoretical predictions for inclusive jet production is a fixed-order perturbative QCD (pQCD) computation. These are available for the inclusive jet production at both NLO and NNLO, and obtained with the NLOJet++ [25, 26] and NNLOJet (rev5918) [27–29] programs, respectively. The NLO computation is implemented in FASTNLO [30], and all calculations are performed assuming massless quarks and five active flavors. The NLO QCD prediction is improved to NLO+NLL with simultaneous jet radius and threshold resummation by applying the k-factors computed in the context of this thesis as

$$k^{\mathrm{NLO+NLL}} = \frac{\sigma^{\mathrm{NLO}} - \sigma^{\mathrm{NLO}}_{\mathrm{sing.}} + \sigma^{\mathrm{NLL}}}{\sigma^{\mathrm{NLO}}}. \tag{7.5}$$

Here $\sigma^{\mathrm{NLO}}_{\mathrm{sing.}}$ stands for the singular terms in the NLO calculation. They are replaced by the resummed σ^{NLL}, which are computed using the NLL- JET calculation, provided by the authors of Ref. [31]. Following their approach, the σ^{NLO} in Eq. (7.5) is calculated with the modified Ellis-Kunszt Soper (MEKS) code version 1.0 [32], in which the $p_{\mathrm{T}}^{\mathrm{max}}$ scale is chosen for the k-factor computation. The predictions for the normalized triple-differential $\mathrm{t\bar{t}}$ cross section are available at NLO, and detailed in Ref. [11].

For the DIS data, the scales μ_r and μ_f are set to the four-momentum transfer Q. For inclusive jet cross section measurements, they are set to the individual jet p_{T}, which was found to be a better scale choice than the leading-jet transverse momentum $p_{\mathrm{T}}^{\mathrm{max}}$ in Ref. [28]. Note however that this scale choice is not possible for the predictions in Refs. [31, 33], for which the $p_{\mathrm{T}}^{\mathrm{max}}$ scale was chosen as discussed above.

The NLO QCD analysis is performed using SM or, alternatively, SMEFT predictions. The latter study incorporates effective corrections corresponding to the color-singlet 4-quark CI described in Sect. 3.2. The SMEFT prediction for the double-differential inclusive jet production cross section is then computed as

$$\sigma^{\mathrm{SMEFT}} = \sigma^{\mathrm{NLO}}_{\mathrm{FASTNLO}} \times k^{\mathrm{NLO+NLL}} \times \mathrm{EW} \times \mathrm{NP} + \mathrm{CI}. \tag{7.6}$$

The factor $k^{\mathrm{NLO+NLL}}$ is given in Eq. (7.5). The utilized EW corrections are at NLO [34]. The contribution from the real production of EW bosons in association with jets is estimated at NLO with MCFM [35–37], and is found to be at most at percent level, which is negligible for the present analysis. However, no uncertainty associated with the EW corrections is yet available. Nonperturbative effects are corrected for by the factor NP, with values corresponding to the average of the corrections obtained with the PYTHIA 8 and HERWIG++ [38] generators. The prediction for the CI contribution is available at NLO, and obtained with the CIJET software [18] interfaced to XFITTER.

The theoretical predictions for the triple-differential $\mathrm{t\bar{t}}$ production cross section at NLO were published with Ref. [11], and calculated with the MADGRAPH5_aMC@NLO (version 2.6.0) [39, 40] framework in the fixed-order mode, interfaced with the AMCFAST (version 1.3.0) [41] and APPLGRID (version 1.4.70) programs. The fixed-order predictions have no variable parameters aside μ_r, μ_f. The APPLGRID tables are produced for fixed values of $m_{\mathrm{t}}^{\mathrm{pole}}$, and the predictions as a function of $m_{\mathrm{t}}^{\mathrm{pole}}$ are obtained by linear interpolation using different $m_{\mathrm{t}}^{\mathrm{pole}}$ values. It was noted in Ref. [11] that there is no significant dependence on the choice of the particular $m_{\mathrm{t}}^{\mathrm{pole}}$ values for the linear interpolation. The cross sections for the $N_{\mathrm{jet}} = 0(1)$ bin are obtained as the difference of the inclusive $\mathrm{t\bar{t}}$($\mathrm{t\bar{t}}$ + 1 jet) and $\mathrm{t\bar{t}}$ + 1 jet ($\mathrm{t\bar{t}}$ + 2 jets) cross sections.

For the triple-differential $t\bar{t}$ cross section, the number of active flavors is set to $n_f = 5$ and the QCD scale choice $\mu_r = \mu_f = \frac{1}{2}\sum_i m_{\mathrm{T}i}$ is used, following Ref. [11]. Here $m_{\mathrm{T}i} \equiv \sqrt{m_i^2 + p_{\mathrm{T}i}^2}$ are transverse masses, computed using the parton masses m_i and transverse momenta $p_{\mathrm{T}i}$, with the sum over i covering the final-state partons t, $\bar{t}$, and at most three light partons in a $t\bar{t} + 2$ jets scenario [11].

7.4 The Simultaneous High-Precision Extraction of PDFs and α_S at NNLO

The CMS 13 TeV inclusive jet production cross section data are used together with the inclusive DIS cross section of HERA in a full QCD analysis at NNLO, in which the PDFs and α_S are extracted simultaneously. The NNLO analysis reported in Ref. [1] is performed using a k-factor approach, and the results are given in Appendix A for reference. However, full NNLO interpolation grids for the double-differential inclusive jet cross section have recently become available [42], and the precision of the analysis presented in this Section is improved by using these grids. The additional flexibility in the parametrization, allowed by inclusion of the CMS data, is investigated in a parametrization scan, described in Sect. 6.5. As a result, the PDFs are parametrized at the starting scale $Q_0^2 = 1.9\,\mathrm{GeV}^2$ as:

$$x\mathrm{g}(x) = A_{\mathrm{g}} x^{B_{\mathrm{g}}} (1-x)^{C_{\mathrm{g}}} (1 + D_{\mathrm{g}} x + E_{\mathrm{g}} x^2), \tag{7.7}$$

$$x\mathrm{u}_v(x) = A_{\mathrm{u}_v} x^{B_{\mathrm{u}_v}} (1-x)^{C_{\mathrm{u}_v}} (1 + E_{\mathrm{u}_v} x^2), \tag{7.8}$$

$$x\mathrm{d}_v(x) = A_{\mathrm{d}_v} x^{B_{\mathrm{d}_v}} (1-x)^{C_{\mathrm{d}_v}}, \tag{7.9}$$

$$x\overline{\mathrm{U}}(x) = A_{\overline{\mathrm{U}}} x^{B_{\overline{\mathrm{U}}}} (1-x)^{C_{\overline{\mathrm{U}}}} (1 + D_{\overline{\mathrm{U}}} x), \tag{7.10}$$

$$x\overline{\mathrm{D}}(x) = A_{\overline{\mathrm{D}}} x^{B_{\overline{\mathrm{D}}}} (1-x)^{C_{\overline{\mathrm{D}}}} (1 + E_{\overline{\mathrm{D}}} x^2). \tag{7.11}$$

In comparison to the HERAPDF2 parametrization, the gluon PDF in Eq. (7.13) does not contain the negative terms $-A'_{\mathrm{g}} x^{B'_{\mathrm{g}}} (1-x)^{C'_{\mathrm{g}}}$, but both the linear term with D_{g}, and the second-order term with E_{g}. Additionally, the parametrization in Eq. (7.11) contains the $E_{\overline{\mathrm{D}}}$ term not included in HERAPDF2.

The PDFs obtained in this analysis, evolved to $\mu_f^2 = 30000\,\mathrm{GeV}^2 \approx m_{\mathrm{t}}^2$, are shown in Fig. 7.2, where the individual contributions of the fit, model and parametrization uncertainties are shown. The strong coupling constant, extracted simultaneously with the PDFs, results in

$$\alpha_S(m_{\mathrm{Z}}) = 0.1166 \pm 0.0014\,(\mathrm{fit}) \pm 0.0007\,(\mathrm{model}) \pm 0.0004\,(\mathrm{scale}) \pm 0.0001\,(\mathrm{param.}), \tag{7.12}$$

which agrees with the previous extractions of $\alpha_S(m_{\mathrm{Z}})$ at NNLO at hadron colliders [43–46].

The impact of the present CMS jet data in a full QCD fit (HERA + CMS fit) is demonstrated by comparing of the resulting PDFs with an alternative fit, where

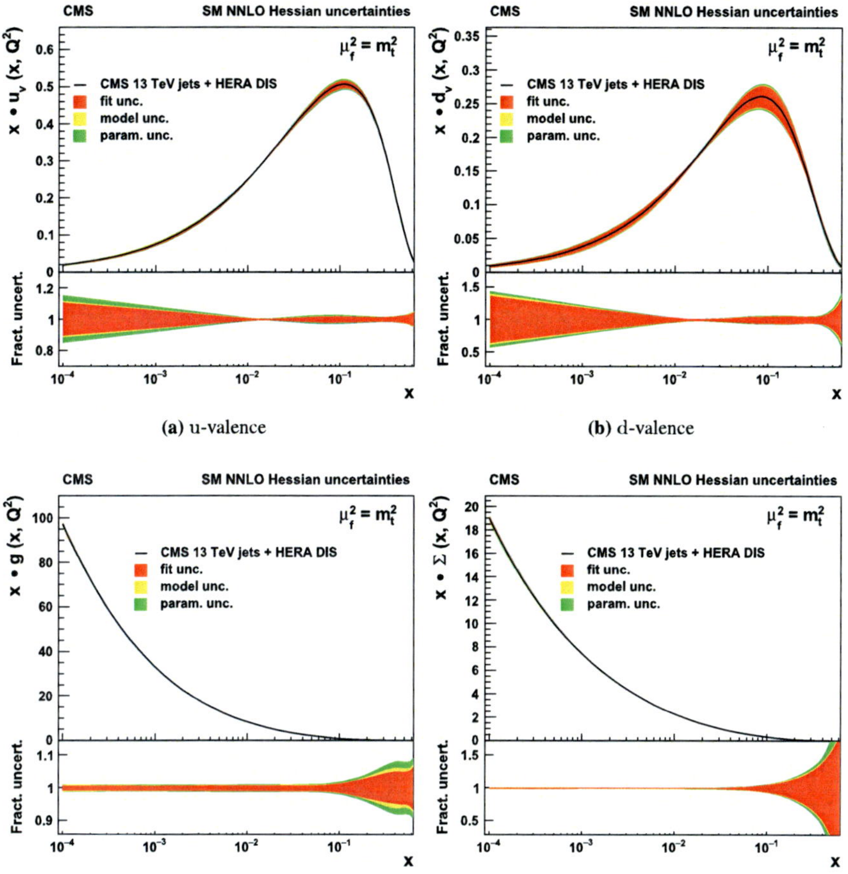

(a) u-valence **(b)** d-valence

(c) Gluon **(d)** Sea

Fig. 7.2 The valence quark, gluon, and sea quark distributions shown as a function of x at the scale $\mu_f^2 = 30000\,\text{GeV}^2 \approx m_t^2$, resulting from the NNLO fit using HERA DIS together with the CMS inclusive jet cross section at $\sqrt{s} = 13\,\text{TeV}$. The inclusive jet cross section is obtained using NNLO interpolation grids. Contributions of fit, model, and parametrization uncertainties for each PDF are shown. In the lower panels, the relative uncertainty contributions are presented [1]

only the HERA data are used (HERA-only fit). Since the inclusive DIS data have much lower sensitivity to the value of $\alpha_S(m_Z)$, it is fixed to the result of the HERA + CMS fit for the HERA-only fit. A comparison of the resulting PDFs is presented in Fig. 7.3. The uncertainty is significantly reduced once the CMS jet measurements are included.

The global and partial χ^2 values for each data set, for the HERA-only and HERA + CMS fits, are listed in Table 7.4, where the χ^2 values illustrate a general agreement among all the data sets. The somewhat high χ^2/N_{dof} values for the combined DIS

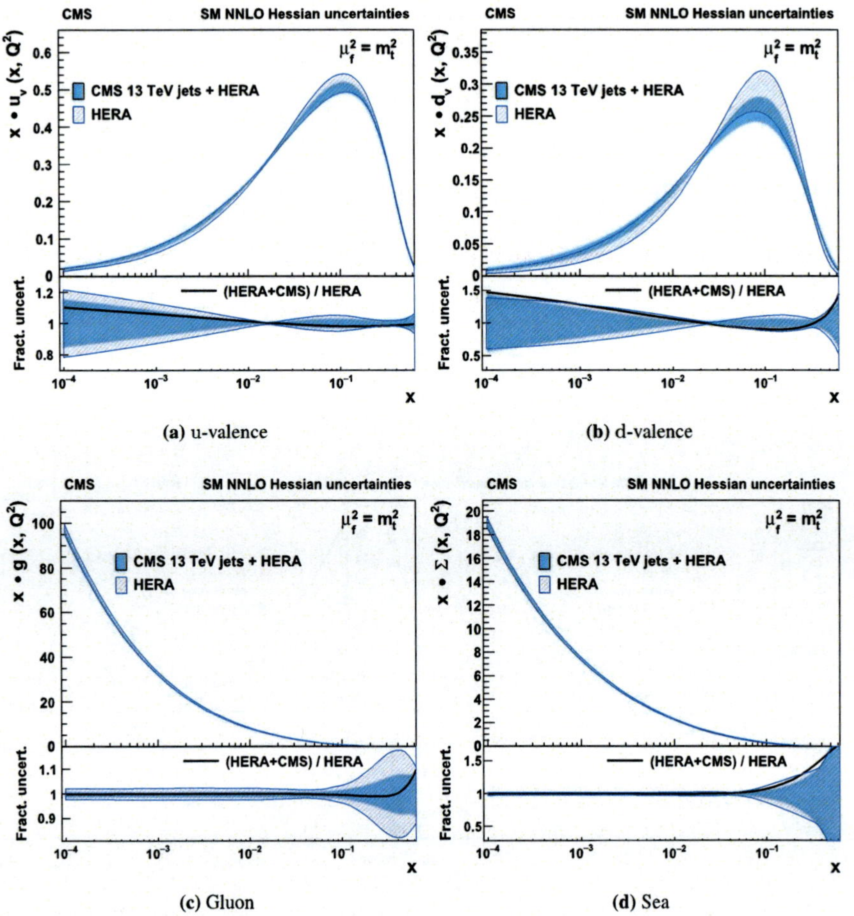

(a) u-valence

(b) d-valence

(c) Gluon

(d) Sea

Fig. 7.3 The valence quark, gluon and sea quark distributions shown as a function of x at the scale $\mu_f^2 \approx m_t^2$. The filled (hatched) band represents the results of the NNLO fit using HERA DIS and the CMS inclusive jet cross section at $\sqrt{s} = 13\,\text{TeV}$ (using the HERA DIS data only). The PDFs are shown with their total uncertainty. The inclusive jet cross section is obtained using NNLO interpolation grids. In the lower panels, the comparison of the relative PDF uncertainties is shown for each distribution. The dashed line corresponds to the ratio of the central PDF values of the two variants of the fit [1]

data are very similar to those observed in Ref. [6], where they are investigated in detail. The correlations between the resulting fit parameters are shown in Fig. 7.4.

A summary of the α_S measurements at hadron colliders and the world average is presented in Fig. 7.5. The presented result is the first extraction of α_S at NNLO using jet production cross section data from pp collisions. The other two CMS analyses performed at NNLO utilize $t\bar{t}$ cross section data at 7 TeV [43], and 13 TeV [44], while

Table 7.4 Partial χ^2 per number of data points N_{dp} and the global χ^2 per degree of freedom, N_{dof}, as obtained in the QCD analysis at NNLO of HERA + CMS jet data and HERA-only data. In the DIS data, the proton beam energy is given as E_p and the electron energy is 27.5 GeV. The HERA data are labeled NC for neutral current and CC for charged current [1]

Data sets		HERA-only Partial χ^2/N_{dp}	HERA + CMS Partial χ^2/N_{dp}
HERA I+II NC	e^+p, $E_p = 920$ GeV	378/332	376/332
HERA I+II NC	e^+p, $E_p = 820$ GeV	60/63	60/63
HERA I+II NC	e^+p, $E_p = 575$ GeV	201/234	202/234
HERA I+II NC	e^+p, $E_p = 460$ GeV	208/187	209/187
HERA I+II NC	e^-p, $E_p = 920$ GeV	223/159	227/159
HERA I+II CC	e^+p, $E_p = 920$ GeV	46/39	46/39
HERA I+II CC	e^-p, $E_p = 920$ GeV	55/42	56/42
CMS jets 13 TeV	$0.0 < \lvert y\rvert < 0.5$		8.6/22
	$0.5 < \lvert y\rvert < 1.0$		23/21
	$1.0 < \lvert y\rvert < 1.5$		13/19
	$1.5 < \lvert y\rvert < 2.0$		14/16
Correlated χ^2		66	81
Global χ^2/N_{dof}		1231/1043	1302/1118

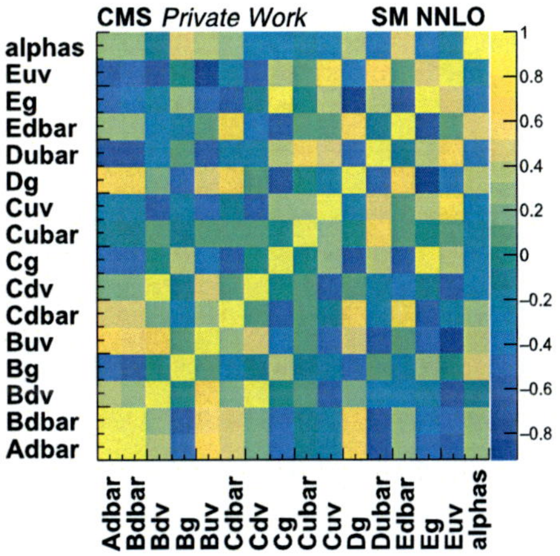

Fig. 7.4 Fit parameter correlations in the NNLO analysis

the H1 analysis of Ref. [45] utilizes inclusive jet and dijet production cross sections from neutral-current DIS data measured from ep collisions at HERA.

The PDFs extracted in the NNLO analysis are compared to the CT18NNLO PDF set [47] in Fig. 7.6. Good agreement between the u_v distributions and uncertainties extracted in this analysis (CMS jet NNLO) and the CT18NNLO PDF is observed. The CT18NNLO PDF has stronger constraints on the d_v and singlet distributions, in particular due to the inclusion of Drell-Yan measurements, which are strongly

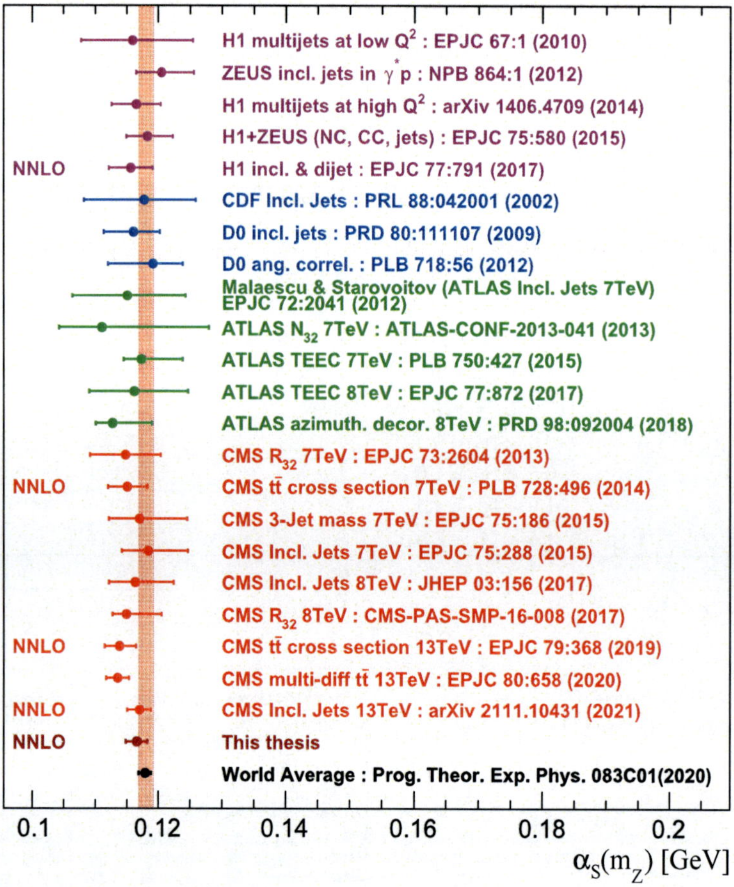

Fig. 7.5 Values of $\alpha_S(m_Z)$ resulting from different measurements, and the world average. The analyses performed at NNLO are indicated separately. Image adapted from Ref. [48], extended to include the NNLO result given in Eq. (7.12)

sensitive to the d_v distribution. Interestingly, the gluon PDF is better constrained in the CMS jet NNLO fit, although the CT18 PDF set includes considerably more data sets, demonstrating the power of the presented jet measurements to constrain the gluon distribution. The gluon and singlet distributions of the CT18 PDF are slightly below the extracted distributions, as expected from the α_S value of 0.118 in the CT18NNLO PDF.

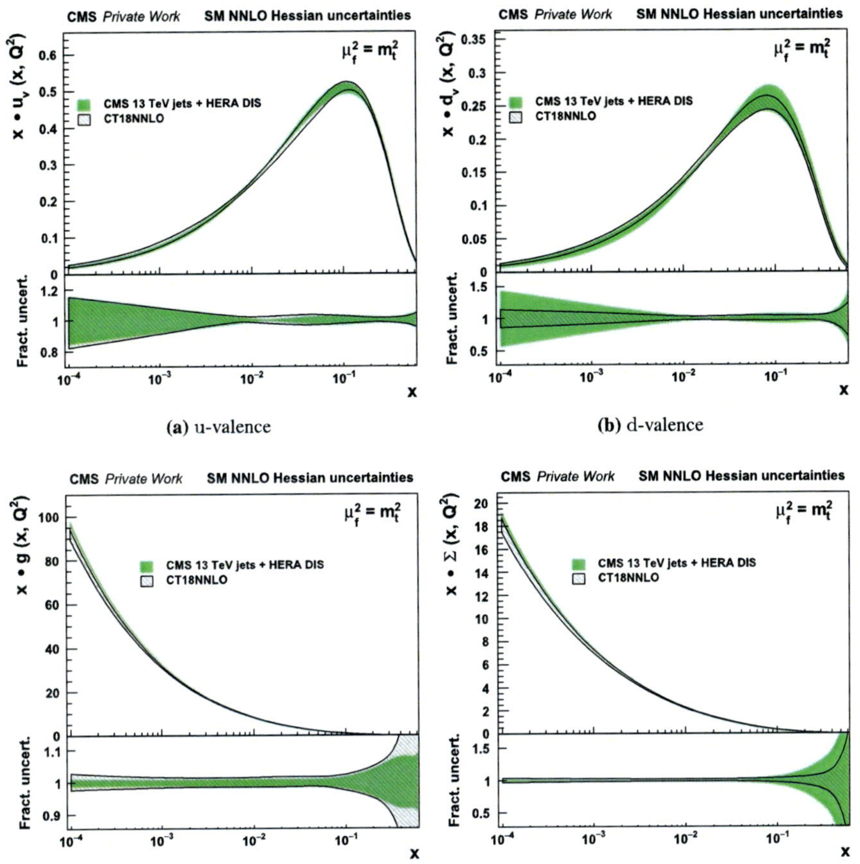

(a) u-valence

(b) d-valence

(c) Gluon

(d) Sea

Fig. 7.6 A comparison of the CT18NNLO valence quark, gluon, and sea quark distributions, shown as a function of x at the scale $\mu_f^2 \approx m_t^2$, and the results of the SM fit at NNLO using the HERA DIS and the CMS inclusive jet cross section data at $\sqrt{s} = 13$ TeV. The green band indicates the total uncertainties for each PDF in the NNLO fit, and the black shaded band shows the CT18NNLO uncertainties converted to 68% CL. The lower panels indicate the relative uncertainty contributions

7.5 Global SMEFT Interpretation of the LHC Data at NLO

In many BSM scenarios the effects of new physics are expected to appear at high-p_T and low-y. However, there is a risk of bias towards results agreeing with the SM, since analyses depend on PDFs extracted from experimental data in the same kinematic region and assuming SM physics. In this thesis, a novel method for an indirect search for 4-quark CI is explored by performing a simultaneous interpretation of the data in terms of the SM and new physics.

The PDFs, $\alpha_S(m_Z)$ and m_t^{pole} are extracted simultaneously with the constraints on BSM physics using the CMS 13 TeV measurements of the inclusive jet cross section and the triple-differential normalized $t\bar{t}$ cross section together with the HERA DIS cross sections in a SMEFT fit. Here, the SM prediction for the inclusive jet cross section is modified to account for the 4-quark CI described in Sect. 3.2. Since the selection of utilized data is different to the analysis in Sect. 7.4, the parametrization is reinvestigated by performing a new parametrization scan, resulting in

$$x\text{g}(x) = A_\text{g} x^{B_\text{g}} (1-x)^{C_\text{g}} (1 + E_\text{g} x^2), \tag{7.13}$$

$$x\text{u}_v(x) = A_{\text{u}_v} x^{B_{\text{u}_v}} (1-x)^{C_{\text{u}_v}} (1 + D_{\text{u}_v} x + E_{\text{u}_v} x^2), \tag{7.14}$$

$$x\text{d}_v(x) = A_{\text{d}_v} x^{B_{\text{d}_v}} (1-x)^{C_{\text{d}_v}} (1 + D_{\text{d}_v} x), \tag{7.15}$$

$$x\overline{\text{U}}(x) = A_{\overline{\text{U}}} x^{B_{\overline{\text{U}}}} (1-x)^{C_{\overline{\text{U}}}}, \tag{7.16}$$

$$x\overline{\text{D}}(x) = A_{\overline{\text{D}}} x^{B_{\overline{\text{D}}}} (1-x)^{C_{\overline{\text{D}}}}. \tag{7.17}$$

In comparison to the HERAPDF2 [6] parametrization, the gluon PDF in Eq. (7.13) contains the second-order term with E_g instead of the additional negative terms $-A'_\text{g} x^{B'_\text{g}} (1-x)^{C'_\text{g}}$. Furthermore, the HERAPDF2 parametrization contains $D_{\overline{\text{U}}}$, but not the D_{u_v} and D_{d_v} terms.

The analysis is performed using SMEFT and, alternatively, SM predictions as a cross-check. The SMEFT interpretation is performed at NLO, to be consistent with the order of the theoretical prediction for the $t\bar{t}$ data and for the CI corrections to the SM Lagrangian. The Wilson coefficient c_1 is introduced as a free parameter in the fit, assuming different values for the scale of the new interaction $\Lambda \in \{5, 10, 13, 20, 50\}$ TeV. For the vector-like and axial vector-like models, the coefficients c_3 and c_5 are obtained from c_1 by the relations given in Table 6.1. Independent of the value of Λ, the strong coupling constant and the top quark mass in the SMEFT fits result in $\alpha_S(m_Z) = 0.1187 \pm 0.0016\,(\text{fit}) \pm 0.0005\,(\text{model}) \pm 0.0023\,(\text{scale}) \pm 0.0018\,(\text{param.})$, and $m_t^{\text{pole}} = 170.4 \pm 0.6\,(\text{fit}) \pm 0.1\,(\text{model}) \pm 0.1\,(\text{scale}) \pm 0.2\,(\text{param.})$ GeV. Notably, the scale uncertainty in α_S is significantly larger than for the NNLO result in Eq. (7.12).

The PDFs resulting from the SMEFT fits at different values of Λ and for different CI models agree with each other and with the PDFs in the SM fit, as illustrated in Fig. 7.7, in which the PDFs obtained in the SMEFT fit using the left-handed CI model, and in the SM fit, are compared. The PDFs are shown only with their fit uncertainty obtained by using the Hessian method.

To account for possible non-Gaussian tails, the PDF uncertainties are alternatively obtained by using the MC method explained in Sect. 6.3, based on 800 replicas. The Hessian and the MC uncertainties in the SMEFT fit are compared in Fig. 7.8. The uncertainties obtained by using the MC method are larger at high x, which might suggest non-Gaussian tails in the PDF uncertainties. However this is not reflected in the uncertainty in c_1 coefficients; the respective uncertainties obtained by Hessian or MC methods agree well.

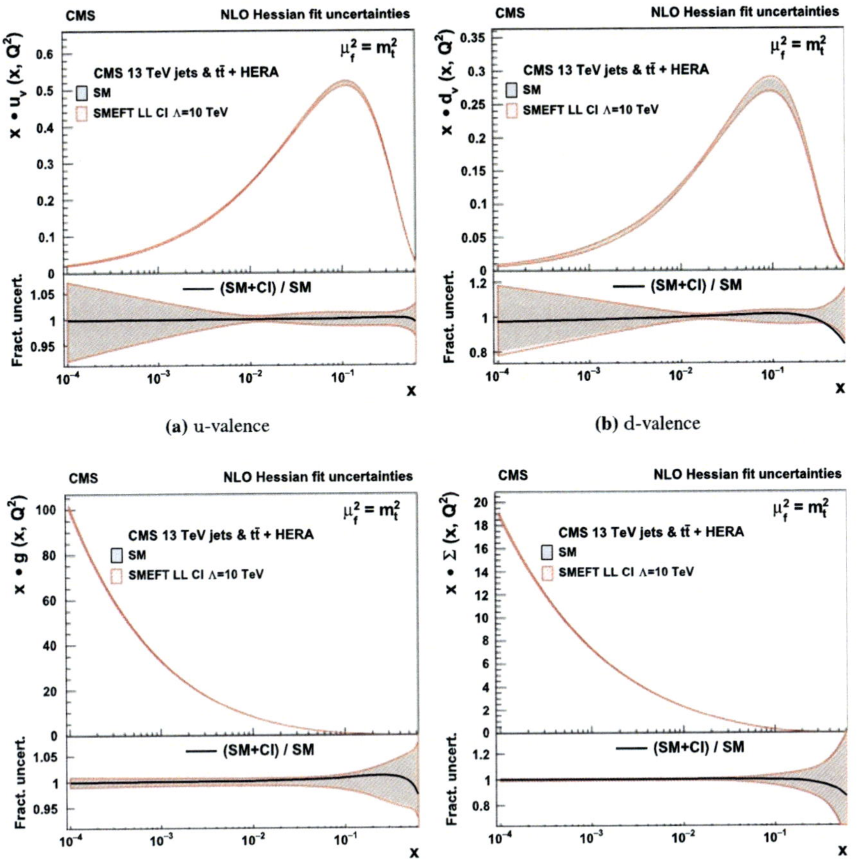

(a) u-valence

(b) d-valence

(c) Gluon

(d) Sea

Fig. 7.7 A comparison of the valence quark, gluon, and sea quark distributions shown as a function of x, at the scale $\mu_f^2 \approx m_t^2$, resulting from the SMEFT and SM fits using HERA DIS together with the CMS inclusive jet cross section and the normalized triple-differential cross section of $t\bar{t}$ production at $\sqrt{s} = 13$ TeV. The SMEFT fit result shown here is attained using the left-handed CI model with $\Lambda = 10$ TeV; the results using other CI models and Λ values are very similar [1]

The Wilson coefficients c_1 are obtained for different assumptions on the value of Λ, as listed in Table 7.5. All SMEFT fits lead to negative c_1, which translates into positive interference with the SM gluon exchange. Nonetheless, there is no statistically significant difference to the SM, for which $c_1 = 0$. The ratio c_1/Λ^2 is illustrated for $\Lambda = 50$ TeV in Fig. 7.9 and is observed to remain constant for various values of Λ.

Conventional searches for CI fix the values of Wilson coefficient to $+1$ (-1) for a destructive (constructive) interference with the SM gluon exchange, and impose exclusion limits on the scale Λ [49]. In the present analysis, all the fitted Wilson

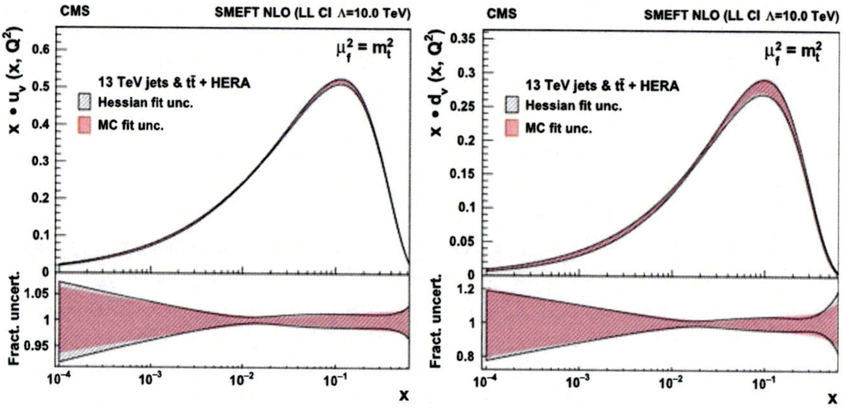

(a) u-valence **(b)** d-valence

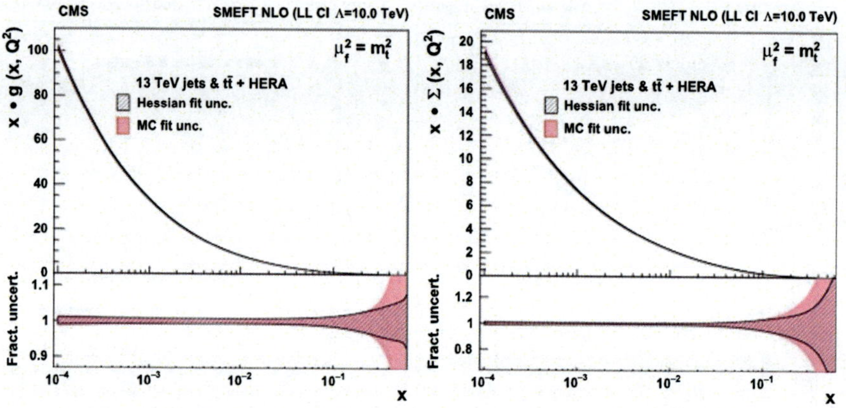

(c) Gluon **(d)** Sea

Fig. 7.8 The valence quark, gluon, and sea quark distributions shown as a function of x, at the scale $\mu_f^2 \approx m_t^2$, with a comparison of the fit uncertainties obtained using the Hessian (hashed) and Monte Carlo (solid) methods. The PDFs result from the SMEFT fit, using the left-handed CI model with $\Lambda = 10\,\text{TeV}$ [1]

coefficients are negative, with c_1 close to -1 for $\Lambda = 50\,\text{TeV}$. These results are translated into 95% CL exclusion limits for models with constructive interference by using the equation $\Delta\chi^2(c_1') = (c_1' - c_1)^2/\sigma_{\text{tot}}^2$, where c_1 are the Wilson coefficients resulting from the fit and σ_{tot}^2 are their total uncertainties, while the SM reference value is obtained as $\Delta\chi^2(0)$. The 95% CL exclusion limit Λ' for the CI scale is found by solving for c_1' in $\Delta\chi^2(c_1') = \Delta\chi^2(0) + 4$ and inserting to $\Lambda' = \Lambda/\sqrt{|c_1'|}$. The resulting limits are $\Lambda > 24\,\text{TeV}$ for the purely left-handed, $\Lambda > 32\,\text{TeV}$ for the vector-like and $\Lambda > 31\,\text{TeV}$ for the axial-vector-like CI model. The most stringent comparable result is obtained in the analysis of dijet cross section at $\sqrt{s} = 13\,\text{TeV}$ by the ATLAS Collaboration [50], in which the 95% CL exclusion limits for purely

Table 7.5 The Wilson coefficients c_1 resulting from fits with various values of the scale Λ. For the vector-like and axial vector-like models, c_3 and c_5 are obtained from c_1 as indicated in Table 6.1. The fit uncertainties are obtained with the Hessian method [1]

Scale	CI model	c_1	Fit	Model	Scale	Param.
$\Lambda = 5\,\text{TeV}$	Left-handed	−0.017	0.0047	0.0001	0.004	0.002
	Vector-like	−0.009	0.0026	0.0001	0.002	0.001
	Axial vector-like	−0.009	0.0025	0.0001	0.002	0.001
$\Lambda = 10\,\text{TeV}$	Left-handed	−0.068	0.019	0.003	0.016	0.009
	Vector-like	−0.037	0.011	0.002	0.008	0.006
	Axial vector-like	−0.036	0.011	0.003	0.008	0.005
$\Lambda = 13\,\text{TeV}$	Left-handed	−0.116	0.033	0.006	0.026	0.015
	Vector-like	−0.063	0.018	0.004	0.015	0.008
	Axial vector-like	−0.062	0.018	0.003	0.014	0.008
$\Lambda = 20\,\text{TeV}$	Left-handed	−0.28	0.08	0.01	0.06	0.04
	Vector-like	−0.15	0.04	0.01	0.04	0.02
	Axial vector-like	−0.15	0.04	0.01	0.04	0.02
$\Lambda = 50\,\text{TeV}$	Left-handed	−1.8	0.53	0.08	0.42	0.23
	Vector-like	−1.0	0.28	0.05	0.23	0.13
	Axial vector-like	−1.0	0.29	0.04	0.23	0.13

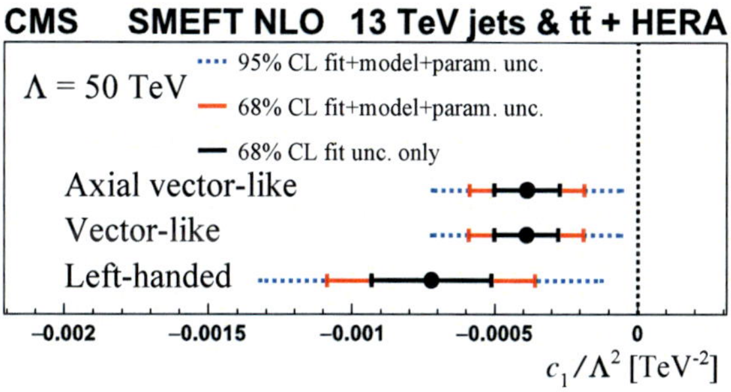

Fig. 7.9 The ratio of the Wilson coefficients c_1 to Λ^2, with $\Lambda = 50\,\text{TeV}$ and the c_1 resulting from the SMEFT fits at NLO. The coefficients c_3 and c_5 are obtained for the vector-like and axial vector-like models via the relations given in Table 6.1. The dashed lines indicate the total uncertainty at 95% CL, while the inner (outer) error bars represent the fit (total) uncertainty at 68% CL [1]

left-handed CI of 22 TeV for constructive interference, and of 30 TeV for destructive interference are obtained.

The correlations between the fit parameter results are shown for the left-handed singlet model at $\Lambda = 10\,\text{TeV}$ in Fig. 7.10. All the CI models considered here result in

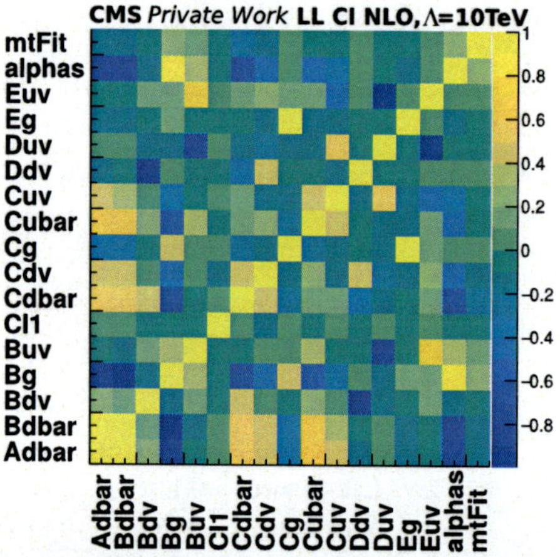

Fig. 7.10 The correlations of the fit parameters in the SMEFT analysis, using the left-handed CI model at $\Lambda = 10\,\text{TeV}$. No significant correlation between the CI Wilson coefficient (CI1) and other parameters is observed

similar correlations. In particular, the fitted Wilson coefficient c_1 is neither strongly correlated nor anticorrelated with any other parameter in the fit.

The PDFs resulting from the SM NLO fit are presented in Fig. 7.11 demonstrating the contributions of the fit, model, and parametrization uncertainties. The NLO values of $\alpha_S(m_Z)$ and of m_t^{pole} determined simultaneously with the PDFs result in $\alpha_S(m_Z) = 0.1188 \pm 0.0017\,(\text{fit}) \pm 0.0004\,(\text{model}) \pm 0.0025\,(\text{scale}) \pm 0.0001\,(\text{param.})$, and $m_t^{\text{pole}} = 170.4 \pm 0.6\,(\text{fit}) \pm 0.1\,(\text{model}) \pm 0.1\,(\text{scale}) \pm 0.1\,(\text{param.})\,\text{GeV}$. These values agree well with those obtained in the SMEFT fit, although the latter have larger parametrization uncertainties due to the increased flexibility in the SMEFT fit. Furthermore, the results are consistent with earlier CMS results [51] and Ref. [11], respectively. The uncertainty in the value of $\alpha_S(m_Z)$ is dominated by the scale variation.

The partial and global χ^2 values for the SMEFT and SM fits are listed in Table 7.6. The fits with all CI models and various Λ values resulted in very similar χ^2 values. In all SMEFT fits, the χ^2 is about 10 units below that of the SM fit, with just the addition of c_1 as an additional free parameter. However, the resulting values of the Wilson coefficient c_1 are consistent with zero within uncertainties for all of the investigated CI models, demonstrating a good description of the data by the SM. Since the SMEFT computation is applied only to the inclusive jet production cross section, the c_1 results are independent of the inclusion of $t\bar{t}$ data. Once the relevant calculations for these data become available, a more global SMEFT interpretation would become possible.

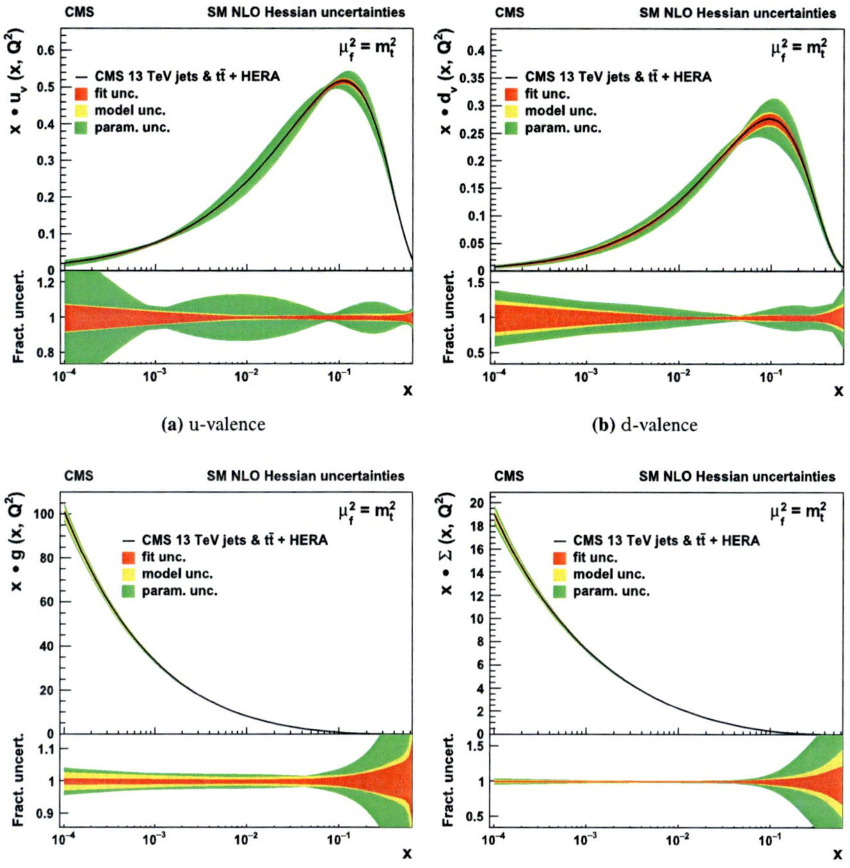

(a) u-valence

(b) d-valence

(c) Gluon

(d) Sea

Fig. 7.11 The valence quark, gluon, and sea quark distributions as functions of x, at the scale $\mu_f^2 \approx m_t^2$, resulting from the SM fit using HERA DIS together with the CMS inclusive jet cross section and the normalized triple-differential cross section of $t\bar{t}$ production at $\sqrt{s} = 13$ TeV. Contributions of fit, model, and parametrization uncertainties for each PDF are shown. In the lower panels, the relative uncertainty contributions are presented [1]

7.6 Profiling CT14 PDF: Impact of the CMS Data

The global PDF set chosen here for the profiling studies is the CT14 PDF [5]. The set is used at NLO or NNLO, respective to the order used in the profiling computations. The set was chosen since neither jet nor $t\bar{t}$ data at 13 TeV were used in its determination, whereas the HERA DIS data are already incorporated into the PDF. Since the triple-differential $t\bar{t}$ cross section has theoretical predictions available only at NLO, they are not considered in NNLO profiling.

Table 7.6 Partial χ^2 per number of data points N_{dp} and the global χ^2 per degree of freedom, N_{dof}, as obtained in the QCD analysis of HERA DIS data and the CMS measurements of inclusive jet production and the normalized triple-differential $t\bar{t}$ production at $\sqrt{s} = 13$ TeV, obtained in SM and SMEFT analyses. The HERA data are labeled NC for neutral current and CC for charged current [1]

Data sets		SMEFT fit Partial χ^2/N_{dp}	SM fit Partial χ^2/N_{dp}
HERA I+II NC	e^+p, $E_p = 920$ GeV	404/332	402/332
HERA I+II NC	e^+p, $E_p = 820$ GeV	60/63	60/63
HERA I+II NC	e^+p, $E_p = 575$ GeV	198/234	198/234
HERA I+II NC	e^+p, $E_p = 460$ GeV	208/187	208/187
HERA I+II NC	e^-p, $E_p = 920$ GeV	223/159	223/159
HERA I+II CC	e^+p, $E_p = 920$ GeV	46/39	46/39
HERA I+II CC	e^-p, $E_p = 920$ GeV	54/42	55/42
CMS 13 TeV $t\bar{t}$ 3D		23/23	23/23
CMS jets 13 TeV	$0.0 < \|y\| < 0.5$	20/22	13/22
	$0.5 < \|y\| < 1.0$	27/21	28/21
	$1.0 < \|y\| < 1.5$	11/19	13/19
	$1.5 < \|y\| < 2.0$	28/16	33/16
Correlated χ^2		115	121
Global χ^2/N_{dof}		1401/1140	1411/1141

The uncertainties considered in the profiling procedure are the PDF and scale uncertainties. The **PDF uncertainty** originates from evaluating the eigenvectors of the CT14 PDF set, and comparing them to the central value, and scaling the uncertainties to 68% CL. The **Scale uncertainty** is computed using the 7-point variation described in Sect. 7.2.1. The total uncertainty is obtained by adding these two uncertainties in quadrature.

Simultaneous profiling of PDF and non-PDF parameters, such as $\alpha_S(m_Z)$, m_t^{pole} or c_1, is not supported in XFITTER, yet. The non-PDF parameters are therefore profiled individually, and $\alpha_S(m_Z)$ is fixed to the CT14 central value of 0.118 when profiling the PDF parameters. Likewise, m_t^{pole} is set to the result of Ref. [11], 170.5 GeV, when using the $t\bar{t}$ cross section, unless the profiled parameter is m_t^{pole}.

The extraction of $\alpha_S(m_Z)$ makes use of the CT14 PDF α_S series at NLO and NNLO, which ranges from 0.1110 to 0.1220, and the profiling is performed individually for each PDF member in the series. The optimal $\alpha_S(m_Z)$ value and its uncertainty are then obtained from a parabolic fit to the χ^2 values. Although any $\alpha_S(m_Z)$ dependence of the k-factors cannot be accounted for, this effect can be considered small due to cancellations in the computation of the k-factor. The scale uncertainty is estimated by varying μ_r and μ_f in the theoretical predictions for the jet cross section and performing individual χ^2 scans for each combination in the 7-point variation procedure.

The results of profiling with the inclusive jet data at NLO are shown in Fig. 7.12, and at NNLO in Fig. 7.13. The additional sensitivity of the 13 TeV jet data is observed to significantly improve the uncertainty in the gluon PDF in the full x range and the sea quark PDF uncertainty at medium x, whereas the valence quark PDF uncertainties remain unchanged.

The χ^2 resulting for the different α_S variations used in the extraction of α_S are depicted in Fig. 7.14. The extracted values read

$$\alpha_S(m_Z) = 0.1170 \pm 0.0018\,(\text{PDF}) \pm 0.0035\,(\text{scale}) \tag{7.18}$$

at NLO and

$$\alpha_S(m_Z) = 0.1128 \pm 0.0016\,(\text{PDF}) \pm 0.0007\,(\text{scale}) \tag{7.19}$$

at NNLO. The NLO result agrees with the world average within uncertainties [49].

The individual impact of the triple-differential CMS $\text{t}\bar{\text{t}}$ cross section of Ref. [11] is also assessed through the profiling studies. Consistent with the available theoretical prediction for the $\text{t}\bar{\text{t}}$ measurements, this analysis is performed at NLO. The improvement in the gluon PDF uncertainty and the result of the α_S scan are shown in Fig. 7.15. The NLO profiling studies of the $\text{t}\bar{\text{t}}$ data result in $m_\text{t}^\text{pole} = 170.6 \pm 0.5\,(\text{PDF}) \pm 0.2\,\text{GeV}\,(\text{scale})$ and $\alpha_S(m_Z) = 0.1127 \pm 0.0014\,(\text{PDF}) \pm 0.0007\,(\text{scale})$. As also observed in [11], the $\text{t}\bar{\text{t}}$ data favor a small value of $\alpha_S(m_Z)$.

The NLO profiling analysis is repeated using both the triple-differential $\text{t}\bar{\text{t}}$ cross section and the inclusive jet cross section. The results are shown in Fig. 7.16, where the uncertainty in the profiled gluon distribution is presented in comparison to that of the original CT14 PDF. The reduction of the uncertainty in the gluon distribution at high x is stronger than in the case when only the CMS inclusive jet cross section is used. This is expected from the additional sensitivity of the $\text{t}\bar{\text{t}}$ production to the gluon distribution at high x. Also the $\alpha_S(m_Z)$ scan is shown in Fig. 7.16, now using both CMS data sets. The PDFs become better constrained, and the resulting NLO value of the strong coupling is $\alpha_S(m_Z) = 0.1154 \pm 0.0009\,(\text{PDF}) \pm 0.0015\,(\text{scale})$, consistent with the result of Ref. [11]. The additional sensitivity of the $\text{t}\bar{\text{t}}$ production to the strong coupling becomes visible in the reduced PDF uncertainties. The profiled pole mass of the top quark results in $m_\text{t}^\text{pole} = 170.3 \pm 0.5\,(\text{PDF}) + 0.2\,(\text{scale})\,\text{GeV}$, consistent with the value obtained in Ref. [11].

The profiling analysis is repeated assuming the SMEFT prediction for the inclusive jet production cross section. While the profiled PDFs remain unchanged with respect to the SM results, the Wilson coefficient c_1 is profiled, assuming the value of the scale of the new interaction $\Lambda = 10\,\text{TeV}$. The results are shown in Fig. 7.17.

As already observed in Sect. 7.5, the results of the profiling studies indicate agreement between the measurements and SM predictions. A direct quantitative comparison of the profiling analysis and the full fit results is however not possible, since the parametrization and the PDF uncertainties differ. With respect to the profiling, the advantage of the full fit is in the properly considered, and therefore mitigated,

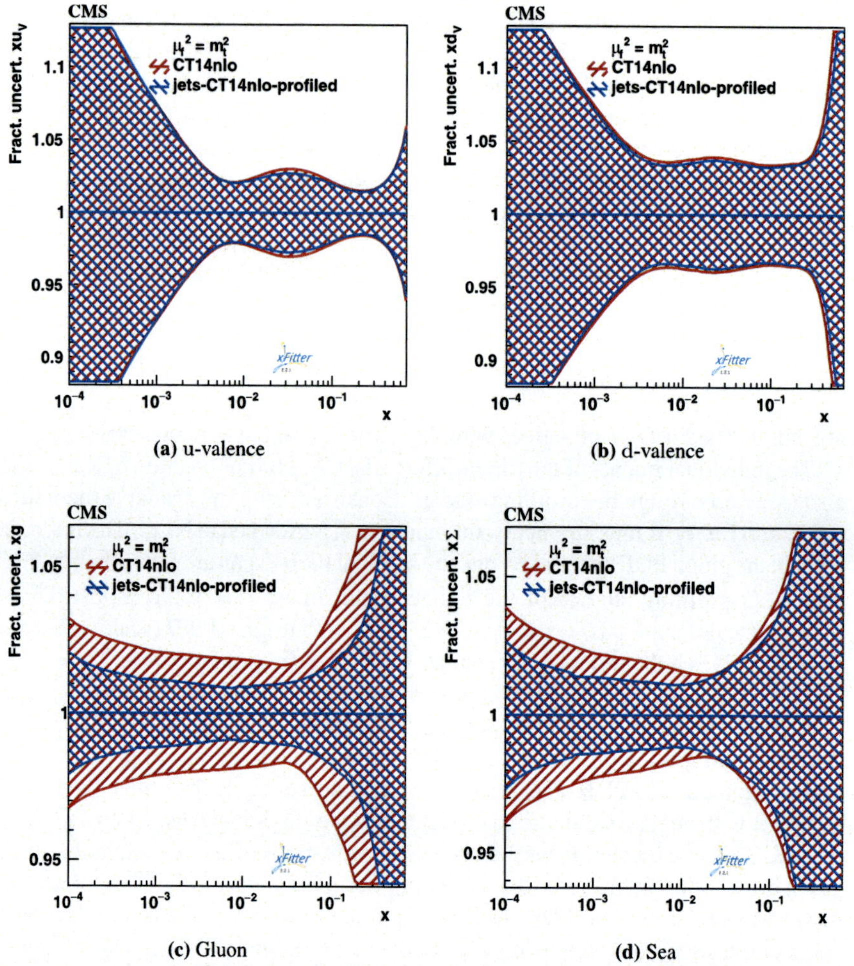

Fig. 7.12 Fractional uncertainties of the valence quark, gluon, and sea quark distributions shown as a function of x, at the scale $\mu_f^2 \approx m_t^2$. The profiling is performed at NLO with the CT14nlo PDF, and including the CMS inclusive jet cross section data at $\sqrt{s} = 13$ TeV, for which the theoretical prediction is at NLO+NLL. The original uncertainty of CT14 is shown in red and the result of PDF profiling in blue [1]

correlations between the QCD parameters and the PDFs. The full SMEFT fit assures that the possible BSM effects are not absorbed in the PDFs and in turn, into the SM prediction, which is the basis for the search for new physics.

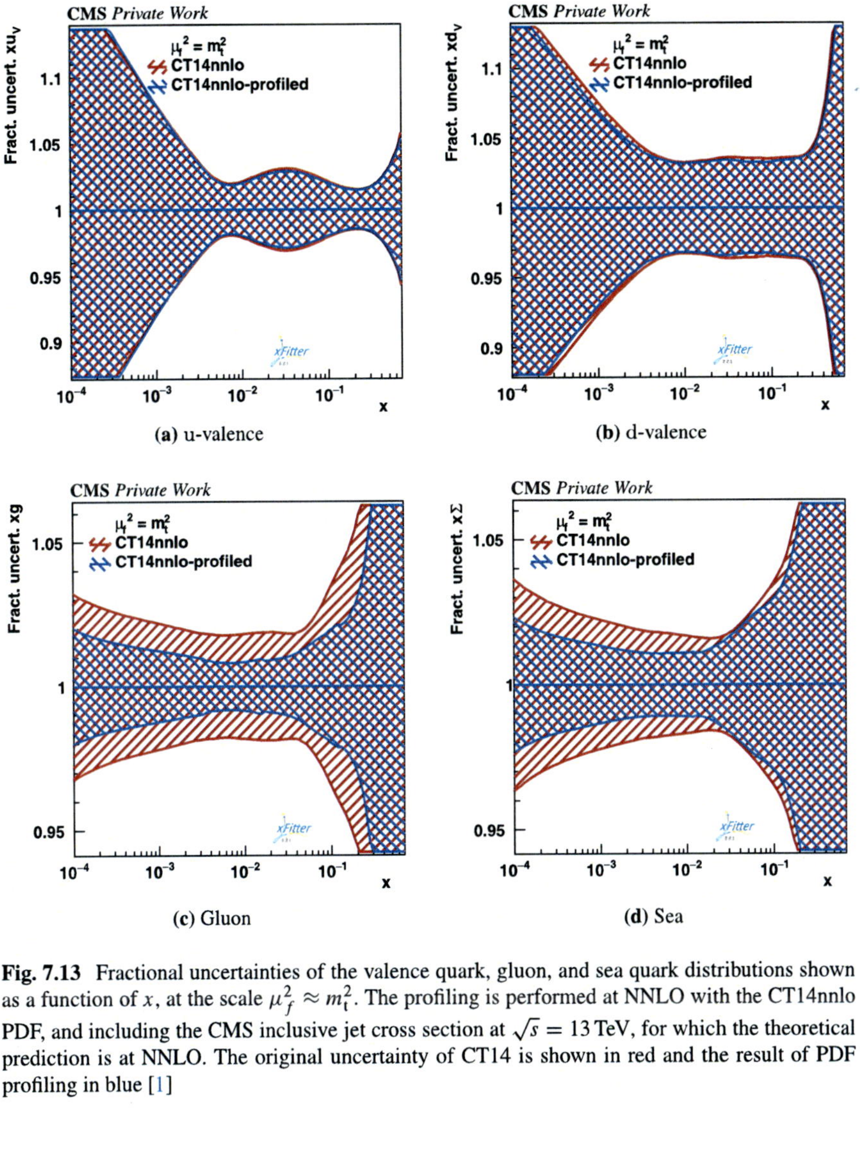

(a) u-valence (b) d-valence (c) Gluon (d) Sea

Fig. 7.13 Fractional uncertainties of the valence quark, gluon, and sea quark distributions shown as a function of x, at the scale $\mu_f^2 \approx m_t^2$. The profiling is performed at NNLO with the CT14nnlo PDF, and including the CMS inclusive jet cross section at $\sqrt{s} = 13\,\text{TeV}$, for which the theoretical prediction is at NNLO. The original uncertainty of CT14 is shown in red and the result of PDF profiling in blue [1]

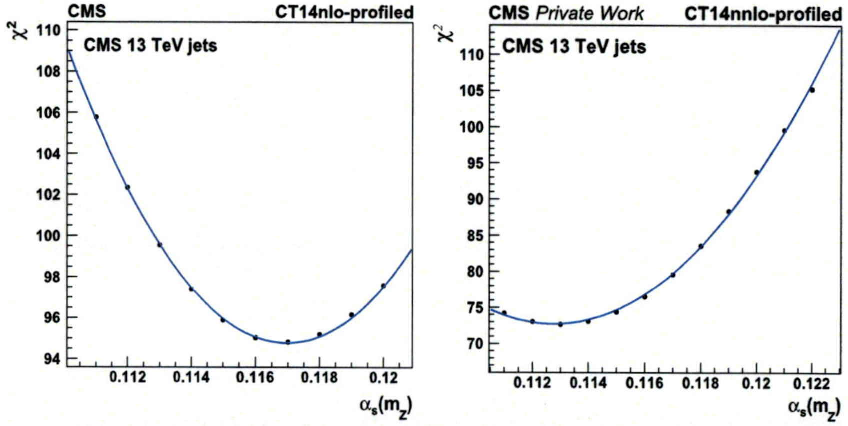

Fig. 7.14 The χ^2 resulting from profiling α_S using the CMS inclusive jet cross section at $\sqrt{s} =$ 13 TeV. *Left*: profiling at NLO using the CT14nlo $\alpha_S(m_Z)$ series, implying NLO+NLL predictions for inclusive jet production. *Right*: profiling at NNLO using the CT14nnlo $\alpha_S(m_Z)$ series and NNLO predictions [1]

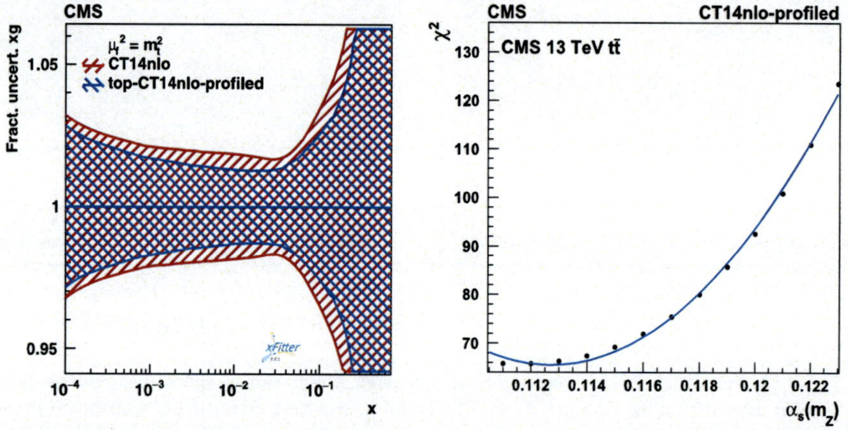

Fig. 7.15 *Left*: fractional uncertainty in the gluon PDF shown as a function of x, at the scale $\mu_f^2 \approx m_t^2$. The profiling is performed at NLO, using the CT14nlo PDF. The included data are the CMS triple-differential $t\bar{t}$ cross sections at $\sqrt{s} = 13$ TeV. The original uncertainty of CT14 is shown in red, and the result of PDF profiling in blue. *Right*: The χ^2 resulting from profiling α_S using the CT14 PDF $\alpha_S(m_Z)$ series. The utilized data sets are the same data as for the left plot

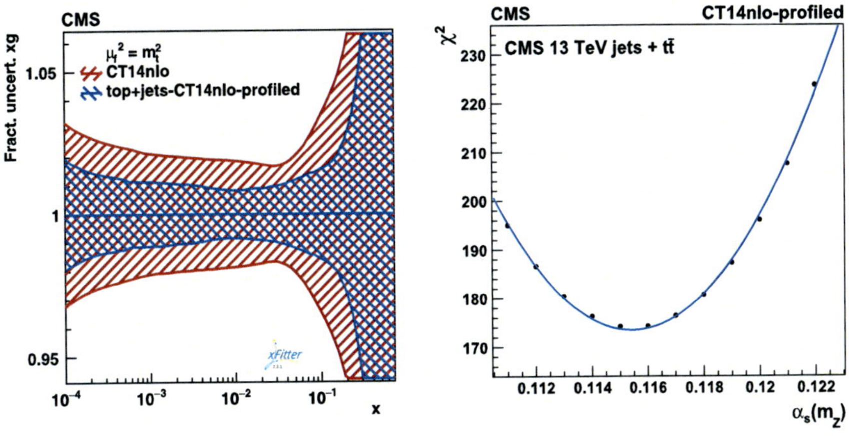

Fig. 7.16 *Left*: fractional uncertainty in the gluon PDF shown as a function of x, at the scale $\mu_f^2 \approx m_t^2$. The profiling is performed at NLO, using the CT14nlo PDF. The included data are the CMS inclusive jet and the triple-differential $t\bar{t}$ cross sections at $\sqrt{s} = 13\,\text{TeV}$. The original uncertainty of CT14 is shown in red, and the result of PDF profiling in blue. *Right*: The χ^2 resulting from profiling α_S using the CT14 PDF $\alpha_S(m_Z)$ series. The utilized data sets are the same data as for the left plot [1]

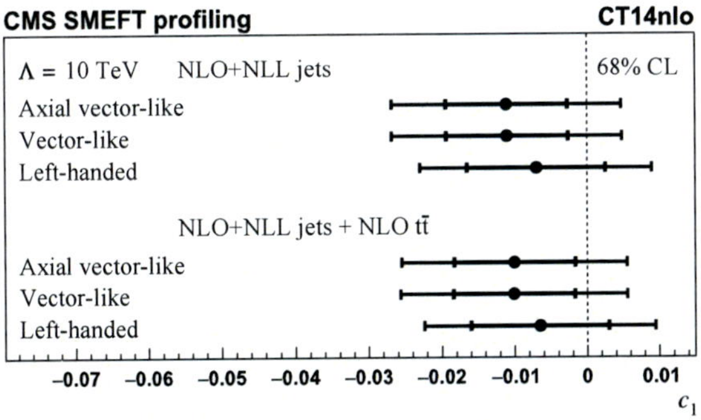

Fig. 7.17 The profiled Wilson coefficient c_1 for the contact interaction models listed in Table 6.1, assuming the left-handed, vector-like, and axial vector-like scenarios, as obtained in the profiling analysis using NLO+NLL calculation for the jet production and the CT14nlo PDF set. The value of $\Lambda = 10\,\text{TeV}$ is assumed. The results are obtained using the CMS measurements of inclusive jet cross section and of normalized triple-differential $t\bar{t}$ cross section at $\sqrt{s} = 13\,\text{TeV}$. The inner error bar shows the PDF uncertainty at 68% CL, while the outer error bar represents the total uncertainty, obtained from the PDF and scale uncertainties, added in quadrature [1]

References

1. Tumasyan A et al (2022) Measurement and QCD analysis of double-differential inclusive jet cross sections in pp collisions at $\sqrt{s} = 13$ TeV. JHEP 02:142
2. Paukkunen H, Zurita P (2014) PDF reweighting in the Hessian matrix approach. JHEP 12:100
3. Schmidt C, Pumplin J, Yuan CP, Yuan P (2018) Updating and optimizing error parton distribution function sets in the Hessian approach. Phys Rev D 98:094005
4. Camarda S et al (2015) QCD analysis of W- and Z-boson production at Tevatron. Eur Phys J C 75:458
5. Dulat S et al (2016) New parton distribution functions from a global analysis of quantum chromody namics. Phys Rev D 93:033006
6. H1 and ZEUS Collaborations (2015) Combination of measurements of inclusive deep inelastic $e^{\pm}p$ scattering cross sections and QCD analysis of HERA data. Eur Phys J C 75:580
7. Cacciari M, Salam GP, Soyez G (2008) The anti-kT jet clustering algorithm. JHEP 04:063
8. Cacciari M, Forte S, Napoletano D, Soyez G, Stagnitto G (2019) Single-jet inclusive cross section and its definition. Phys Rev D 100:114015
9. Dasgupta M, Dreyer FA, Salam GP, Soyez G (2016) Inclusive jet spectrum for small-radius jets. JHEP 06:057
10. Bellm J et al (2020) Jet cross sections at the LHC and the quest for higher precision. Eur Phys J C 80:93
11. Sirunyan AM et al (2020) Measurement of tt normalised multi-differential cross sections in pp colli sions at $\sqrt{s} = 13$ TeV, and simultaneous determination of the strong coupling strength, top quark pole mass, and parton distribution functions. Eur Phys J C 80:658
12. Sirunyan AM et al (2017) Measurement of double-differential cross sections for top quark pair pro duction in pp collisions at $\sqrt{s} = 8$ TeV and impact on parton distribution functions. Eur Phys J C 77:459
13. Sjostrand T, Mrenna S, Skands PZ (2006) PYTHIA 6.4 physics and manual. JHEP 05:026
14. Alekhin S et al (2015) HERAFitter, open source QCD fit project. Eur Phys J C 75:304
15. Bertone V, Botje M, Britzger D et al (2018) XFITTER 2.0.0: an open source QCD fit framework. PoS DIS2017:203
16. Abdolmaleki H et al (2022) xFitter: an open source qcd analysis framework. A resource and reference document for the Snowmass study in (June 2022). arXiv: 2206.12465 [hep-ph]
17. XFitter Developers' Team. https://www.xfitter.org/xFitter/
18. Gao J (2013) CIJET: a program for computation of jet cross sections induced by quark contact inter actions at hadron colliders. Comput Phys Commun 184:2362
19. Botje M (2011) QCDNUM: fast QCD evolution and convolution. Comput Phys Commun 182:490
20. Thorne RS, Roberts RG (1998) An Ordered analysis of heavy flavor production in deep inelastic scattering. Phys Rev D 57:6871
21. Thorne RS (2006) A variable-flavor number scheme for NNLO. Phys Rev D 73:054019
22. Thorne RS (2012) Effect of changes of variable flavor number scheme on parton distribution functions and predicted cross sections. Phys Rev D 86:074017
23. H1 and ZEUS Collaborations (2010) Combined measurement and QCD analysis of the inclusive $e^{\pm}p$ scattering cross sections at HERA. JHEP 01:109
24. Pumplin J et al (2001) Uncertainties of predictions from parton distribution functions. II. The Hessian method. Phys Rev D 65:014013
25. Nagy Z (2002) Three jet cross-sections in hadron-hadron collisions at next-to-leading order. Phys Rev Lett 88:122003
26. Nagy Z (2003) Next-to-leading order calculation of three jet observables in hadron-hadron collision. Phys Rev D 68:094002
27. Currie J, Glover EWN, Pires J (2017) Next-to-next-to leading order QCD predictions for single jet inclusive production at the LHC. Phys Rev Lett 118:072002
28. Currie J et al (2018) Infrared sensitivity of single jet inclusive production at hadron colliders. JHEP 10:155

29. Gehrmann T et al (2018) Jet cross sections and transverse momentum distributions with NNLO-JET. In: Hoang A, Schneider C (eds) PoS RADCOR2017, p 074
30. Britzger D, Rabbertz K, Stober F, Wobisch M (2012) New features in version 2 of the FASTNLO project in 20th international workshop on deep-inelastic scattering and related subjects, 217. https://doi.org/10.3204/DESY-PROC-2012-02/165. arXiv:1208.3641 [hep-ph]
31. Liu X, Moch S-O, Ringer F (2018) Phenomenology of single-inclusive jet production with jet radius and threshold resummation. Phys Rev D 97:056026
32. Gao J et al (2013) MEKS: a program for computation of inclusive jet cross sections at hadron colliders. Comput Phys Commun 184:1626
33. Liu X, Moch S-O, Ringer F (2017) Threshold and jet radius joint resummation for single-inclusive jet production. Phys Rev Lett 119:212001
34. Dittmaier S, Huss A, Speckner C (2012) Weak radiative corrections to dijet production at hadron colliders. JHEP 11:095
35. Campbell JM, Ellis RK (1999) An Update on vector boson pair production at hadron colliders. Phys Rev D 60:113006
36. Campbell JM, Ellis RK, Giele WT (2015) A multi-threaded version of MCFM. Eur Phys J C 75:246
37. Campbell JM, Ellis RK, Williams C (2011) Vector boson pair production at the LHC. JHEP 07:018
38. Bahr M et al (2008) HERWIG++ physics and manual. Eur Phys J C 58:639
39. Alwall J, Herquet M, Maltoni F, Mattelaer O, Stelzer T (2011) MADGRAPH 5: going beyond. JHEP 06:128
40. Alwall J et al (2014) The automated computation of tree-level and next-to-leading order differential cross sections, and their matching to parton shower simulations. JHEP 07:079
41. Bertone V, Frederix R, Frixione S, Rojo J, Sutton M (2014) aMCfast: automation of fast NLO computations for PDF fits. JHEP 08:166
42. Britzger D et al (2022) NNLO interpolation grids for jet production at the LHC. arXiv: 2207.13735 [hep-ph]
43. Chatrchyan S et al (2014) Determination of the top-quark pole mass and strong coupling constant from the tt production cross section in pp collisions at s=7 TeV. Phys Lett B 728:496–517
44. Sirunyan AM et al (2019) Measurement of the tt production cross section, the top quark mass, and the strong coupling constant using dilepton events in pp collisions at $\sqrt{s}$ = 13 TeV. Eur Phys J C 79:368
45. Andreev V et al (2017) Determination of the strong coupling constant S(mZ) in next-to-next-to leading order QCD using H1 jet cross section measurements. Eur Phys J C 77:791
46. Camarda S, Ferrera G, Schott M (2022) Determination of the strong-coupling constant from the Z-boson transverse-momentum distribution. arXiv: 2203.05394 [hep-ph]
47. Hou T-J et al (2021) New CTEQ global analysis of quantum chromodynamics with high-precision data from the LHC. Phys Rev D 103:014013
48. CMS physics results combined Accessed: 2022-01-17. https://twiki.cern.ch/twiki/bin/view/CMSPublic/PhysicsResultsCombined
49. Particle Data Group, Zyla PA et al (2020) Review of particle physics. Prog Theor Exp Phys 2020:083C01
50. Aaboud M et al (2017) Search for new phenomena in dijet events using 37 fb1^- of pp collision data collected at $\sqrt{s}$ = 13 TeV with the ATLAS detector. Phys Rev D 96:052004
51. Khachatryan V et al (2017) Measurement and QCD analysis of double-differential inclusive jet cross sections in pp collisions at $\sqrt{s}$ = 8 TeV and cross section ratios to 2.76 and 7 TeV. JHEP 03:156

Chapter 8
Summary and Conclusions

This thesis presents the most precise measurement of the strong coupling constant α_S at the LHC, and the first examination of the behavior of theoretical predictions in the MSR scheme as a function of the scale R, leading to the first theoretically consistent determination of the top quark MSR mass. The potential of the inclusive jet and top quark-antiquark pair production measurements by the CMS Collaboration is exploited in a SMEFT analysis, where the PDFs, α_S, m_t^{pole} and the couplings of new physics are determined simultaneously. This analysis is the first of its kind at the LHC. The presented investigations are at the cutting edge of jet and top quark phenomenology, and serve as the first step towards a global simultaneous interpretation of the LHC data in terms of the couplings both in both the SM and new physics. Part of the work presented in this thesis has recently been published in *The Journal of High-Energy Physics* [1].

In a SM QCD analysis at NNLO, the double-differential inclusive jet cross section measured by the CMS Collaboration in pp collisions at $\sqrt{s} = 13\,\text{TeV}$ as a function of the jet transverse momentum p_T and the jet rapidity $|y|$, is used together with the HERA measurements of deep inelastic scattering. The value of $\alpha_S(m_Z)$ extracted simultaneously with the parton distributions is $\alpha_S(m_Z) = 0.1166 \pm 0.0017$, which is the most precise result obtained using LHC data, to date. As the precision of the PDFs is often a limiting factor for the precision studies performed using hadron collider data, it is important that experimental collaborations provide the global PDF groups with precise measurements to constrain future global PDF fits. In this regard, the power of the inclusive jet production cross section data is demonstrated by the behavior and accuracy of the PDFs extracted in the analysis, which are comparable with the global CT18 PDF set at NNLO, albeit extracted from a considerably smaller amount of data. The significant improvement in the PDF accuracy obtainable with the 13 TeV inclusive jet data is shown further in a profiling analysis using the CT14 PDF set.

T. Mäkelä, *Towards Global Interpretation of LHC Data*, Springer Theses,
https://doi.org/10.1007/978-3-031-29779-3_8

A further issue stressing the importance of a proper understanding of the PDFs is the possibility for bias in searches for new physics, as they rely on predictions depending on PDFs extracted assuming the SM in the same kinematic region where the effects of new physics are expected to be manifest. The issue is addressed in this thesis by following an unbiased strategy in an analysis at NLO, where the CMS measurement of the normalized triple-differential top quark-antiquark production cross section at $\sqrt{s} = 13\,\text{TeV}$ is used in addition. In this analysis, the SM Lagrangian is modified by introducing SMEFT terms for 4-quark contact interactions, and the Wilson coefficients of the contact interactions are extracted simultaneously with the PDFs, $\alpha_S(m_Z)$ and m_t^{pole}, for various assumed values for the new physics scale Λ. This results in $\alpha_S(m_Z) = 0.1187 \pm 0.0046$, where the uncertainties are dominated by the QCD scale uncertainty, and $m_t^{\text{pole}} = 170.4 \pm 0.8$, and the results for the fitted Wilson coefficients are translated into a 95% confidence level exclusion limits for constructive interference with the SM gluon exchange. The obtained limits correspond to $\Lambda > 24\,\text{TeV}$ for the left-handed model, $\Lambda > 32\,\text{TeV}$ for the vector-like and $\Lambda > 31\,\text{TeV}$ for the axial vector-like CI model. The results are compatible with the SM and the previous limits obtained in conventional searches at the LHC using jet production. The advantage of the approach presented in the thesis is the simultaneous extraction of PDFs, thereby mitigating possible bias in the interpretation of the measurements in terms of physics beyond the SM. As a cross-check, the NLO analysis is repeated using SM predictions. The values of $\alpha_S(m_Z)$ and m_t^{pole} resulting from the analysis are in agreement with those extracted in the SMEFT analysis. The values of $\alpha_S(m_Z)$ resulting from both the SM and SMEFT fits agree with the world average and the previous CMS results using the jet measurements, and the values of m_t^{pole} agree well with the result of the previous CMS analysis using the triple-differential cross section for top quark-antiquark pair production. Although the inclusive jet production is not directly sensitive to m_t^{pole}, the resulting value is improved by the additional constraints on the gluon distribution and $\alpha_S(m_Z)$ provided by the jet measurements.

Furthermore, an implementation of the MSR scheme for the renormalization of the mass of the top quark is presented for the HATHOR inclusive $t\bar{t}$ production cross section calculation at NNLO and the MCFM v6.8 single-differential $t\bar{t}$ production cross section computation at NLO. The evolution of the top quark mass is performed in the MSR scheme for mass scales $R < \overline{m}_t(\overline{m}_t)$, and in the $\overline{\text{MS}}$ scheme for mass scales $\mu_m > \overline{m}_t(\overline{m}_t)$, and allows investigating the behavior of the mass scales independent of the QCD factorization and renormalization scales. Studies of the inclusive $t\bar{t}$ production cross section indicate that μ_r variations dominate the effects of μ_f variations, and that the cross sections computed assuming small values of μ_r and μ_f undergo smaller variations as a function of the mass scale R or μ_m than the predictions computed using high values of μ_r and μ_f. The inclusive cross section also serves as an important cross-check for the single-differential computation, which is further validated by comparing the results of the modified MCFM program to an external computation translating standard pole mass scheme cross section results into the MSR scheme.

The first study of the behavior of the R scale is presented, with particular focus on the $m_{\mathrm{t\bar{t}}}$ distribution. To obtain theoretical predictions that are robust against scale variations, a central value of about 80 GeV is suggested for the scales R, μ_r and μ_f, especially near the $\mathrm{t\bar{t}}$ production threshold. The effects of a lacking treatment for the Coulomb corrections near the $\mathrm{t\bar{t}}$ production threshold are discussed, indicating the importance of their implementation to ensure a proper understanding of the threshold region in future analyses. The investigation culminates in an extraction of the top quark MSR mass from experimental $\mathrm{t\bar{t}}$ pair production cross section data measured as a function of $m_{\mathrm{t\bar{t}}}$. For this, the CMS measurement of the single-differential $\mathrm{t\bar{t}}$ production cross section in pp collisions at $\sqrt{s} = 13\,\mathrm{TeV}$ is utilized. Setting the scales μ_r and μ_f equal to $m_{\mathrm{t}}^{\mathrm{MSR}}(R = 80\,\mathrm{GeV})$ throughout the $m_{\mathrm{t\bar{t}}}$ distribution is observed to yield results consistent with the MC studies presented in Ref. [2], and with Ref. [3] once the extracted MSR mass is translated into the $\overline{\mathrm{MS}}$ mass $\overline{m}_{\mathrm{t}}(\overline{m}_{\mathrm{t}})$. Further, it is observed that the choice of a dynamical scale yielding low μ_r and μ_f values near the $\mathrm{t\bar{t}}$ production threshold reduces the scale uncertainties in the value of the top quark MSR mass extracted from the cross section measured as a function of $m_{\mathrm{t\bar{t}}}$. This results in the value $m_{\mathrm{t}}^{\mathrm{MSRn}}(80\,\mathrm{GeV}) = 169.3^{+0.6}_{-0.7}\,\mathrm{GeV}$, which is thus far the most precise value of the top quark MSRn mass extracted from experimental data.

The author has contributed to the development of the xFITTER open QCD analysis framework. In particular, the computation of 4-quark contact interaction contributions to jet production is possible via an interfaces to CIJET. Moreover, the capabilities of the framework to compute predictions for $\mathrm{t\bar{t}}$ and single top quark production are extended via improvements to the HATHOR interface.

The thesis addresses the precision of the strong coupling constant, the parton distributions, and the top quark mass, as well as the couplings of new physics. The presented studies as well as the software development carried out in the context of the thesis serve to pave the road toward future global interpretations of LHC data.

References

1. Tumasyan A et al (2022) Measurement and QCD analysis of double-differential inclusive jet cross sections in pp collisions at $\sqrt{s}$ = 13 TeV. JHEP 02:142
2. A precise interpretation for the top quark mass parameter in ATLAS Monte Carlo simulation (2021)
3. Sirunyan AM et al (2020) Running of the top quark mass from proton-proton collisions at $\sqrt{s}$ = 13TeV. Phys Lett B 803:135263